WATER POLLUTION

WATER POLLUTION

Edited by

Prof. P.R. TRIVEDI

Chairman

Indian Institute of Ecology and Environment

New Delhi - 110 030

&

Prof. GURDEEP RAJ

Hony. Professor in Environment Science in

I.I.E.E., New Delhi

Reader in Chemistry

D.S. College, University of Delhi,

New Delhi - 110 003

AKASHDEEP PUBLISHING HOUSE

NEW DELHI - 110 002 (INDIA)

AKASHDEEP PUBLISHING HOUSE
4374/4B, Ansari Road, Daryaganj
New Delhi - 110 002
Ph.: 23261597, 23278000
Visit us at: www.anmolpublications.com

Water Pollution

Reprint, 2005

ISBN 81-7158-279-6

PRINTED IN INDIA

Published by Akashdeep Publishing House, New Delhi - 110 002 and
Printed at Mehra Offset Press, Delhi.

Preface

This book entitled "Water Pollution" is a serious attempt to keep pace with continuing rapid developments in the field of water pollution during the last twenty five years. The main aim of this book is to present in a rigorous, quantitative manner many necessary concepts which are essential for solving problems in water pollution. In order to make the book more interesting, data have been carefully selected from developing as well as under developing countries. This book will meet the requirements of a wide range of readers who are involved directly or indirectly with water pollution, including environmental students, research workers, civil and environmental engineers, operators of waste treatment plants, environmental protection agencies, conservationists, law enforcement officials and public administrators.

During the compilation and editing of this book, we have freely consulted the work of numerous authorities, newspapers, weekly magazines and seek the advice of students and colleagues to all of whom the editors are deeply indebted.

Finally the editors express their sincere thanks to the publishers and printer for bringing out this book promptly.

All comments from users on omissions or shortcomings will be highly appreciated.

The Editors

Contents

1

Introduction to Water Pollution

WATER POLLUTION

Introduction

Water pollution is a phenomenon that is characterized by the deterioration of the quality of land water (rivers, lakes, marshes and ground water) or sea water as a result of various human activities. In this chapter, we will first classify various humun activites that cause water pollution, discuss water pollution according to their characteristics, and then take a general view of water pollution which is currently in progress in our modern society. Human activities related with water pollution comprise various industries such as mining, agriculture, stock-breeding, fisheries, forestry, urban human activities, manufacturing industry, construction works and various tertiary industries. Then, in the last section of this chapter, we will classify waste waters according to their physical and chemical properties.

Water pollution is any physical or chemical change in water that can adversely affect organisms. It is a global problem, affecting both the industrialized and the developing nations. The water pollution problems in the rich and the poor nations, however, are quite different in many respects. Heat, toxic metals, acids, sediment, animal and human wastes, and synthetic organic compounds foul the waterways of developed nations. Human and animal wastes, sediment, and pathogenic organisms head the list in the non-industrialized nations. In these countries, unsanitary water and malnutrition account for most of the illness and death.

Water has many different uses, each requiring different levels of prity. Four example, water from the Ohio River may be clean enough to wash steel but may be toxic to fish and wildlife; water suitable for boating or fishing may be unsuitable for swimming; water safe to swin in may be too polluted to drink.

Like air pollutants, water pollutants come from numerous natural and anthropogenic sources. Likewise, water pollutants produced in one nation may flow into others, creating complex international control problems that may take decades to solve.

The term 'water pollution' is referred to the addition to water of an excess of material (or heat) that is harmful to humans, animals, to desire able aquatic life or otherwise causes significant departures from the normal activities of various living communities in or near bodies of water. The National Water Commission stated (1973) that "water gets polluted if it has been not of sufficiently high quality to be suitable for the highest uses people wish to make of it at present on in the future."

In reality, the term water pollution refers to any type of aquatic contamination between two extremes :

1. A highly enriched, over productive biotic community, such as a river or lake with nutrients from sewage or fertilizer (cultural eutrophication), or
2. A body of water poisioned by toxic chemicals which eleminate living organism or even exclude all forms of life.

The expression "water pollution" seems to be clear to all. Nevertheless, it is worth determining its real meaning as this has changed in the course of time. Felfoldy's (1982) precise definition is the following : "Water pollution is every impact which changes the quality of our surface and subsoil waters to such a degree that its suitability either for human consumption or for the support of man's natural life processes with decrease or cease."

Unpolluted *vs.* Polluted Water

Strictly speaking, pollution has been any departure from purity. When environmental pollution has been the topic, the term has come to mean a departure from a normal, rather than from a pure state. This has been particularly true for water. This widely distributed substance has been such a good solvent that it has been never found naturally in a completely pure state.

Even in the most unpolluted geographical areas, rainwaters is having dissolved CO_2, O_2 and N_2 and may also carry in suspension dust or other particulates picked up from the atmosphere. Surface and well waters generally contain dissolved compounds of metals like Na, Mg, Ca, and Fe. The term hard water used to describe water that contains appreciable amounts of such compounds. Even drinking water has been not pure in a chemical sense. Suspended solids have been removed and harmful bacteria destroyed, but many substances still remain in solution. Indeed, absolutely pure water would not be pleasant to drink, because it has been is the impurities that impart water the characteristic "taste" by which it is recognized.

In light of the above facts the term pure, when used in a water pollution context, will mean a state of water in which no substance is present in sufficient concentration to prevent the water from being used for purposes thought of as normal. Normal areas of use include :

1. Recreation and aesthetics
2. Public water supply

3. Fish, other aquatic life, and wildlife
4. Agriculture
5. Industry

Any substance that disallows the normal use of water must be regarded as a water pollutant. Part of the complexity of the water pollution problem arises because the normal uses of water are so varied. Water suited for some used and therefore considered to be unpolluted, may have to be regarded to be polluted when other uses get contemplated.

Water Quality Parameters and Standards

The parameters for water-quality characterization have been given in Table 1.1. The permissible limits, which have been prescribed by the United States Public Health Drinking Water Standards (USPH) and Indian Standards Institution (ISI) have been given for comparision. These refer to domestic water supplies for drinking water (Table 1.1). The ISI values, have been found to be higher than those for USPH.

Surface Waters

The permissible limits of the parameters given above for surface water have been found to be 3-5 times higher. The general characteristics for recreation and aesthetics (swimming, rowing, navigation, etc aquatic life, fish and wildlife, and irrigation have been given as follows :

TABLE 1.1

The Various Parameters Used for Water Quality Characterization and Standards

(Domestic water Supplies)

Parameters	*USPH Standard*	*ISI Standard* (*IS : 2296-1963*)
Physical Characteristics		
Colour, Odour	Colourless, Odourless,	

Taste	Tasteless	
pH	6.0 - 8.5	6.0 – 9.0
Specific conductance	300 μmho cm^{-1}	—
Dissolved oxygen (D.O.)	4.0 – 6.0 (ppm)	3.0
Total dissolved solids	500	—
Suspended solid	5.0	—
Inorganic Chemicals		
Chloride	250	600
Sulphate	250	1000
Cyanide	0.05	0.01
Nitrate + nitrite	<10	—
Fluoride	1.5	3.0
Phosphate	0.1	—
Sulphide	0.1 μ L^{-1} (ppb)	—
Ammonia	0.5	—
Boron	1.0	—
Calcium	100	—
Magnesium	30	—
Arsenic	0.05	0.2
Barium	1.0	—
Cadmium	0.01	—
Chromium (VI)	0.05	0.05
Copper	1.0	—
Iron (filtrable)	<0.3	—
Lead	<0.05	0.1
Manganese (filtrable)	< 0.05	—
Mercury	0.001	—
Selenium	0.01	0.05
Silver	0.05	—
Uranium	5.0	—
Zinc	5.5	—
Organics		
COD	4.0	
Carbon $CHCl_3$	0.15	

Extract (CCE)			
Methylene blue active substances	0.5		
Phenols	0.001		0.005
Pesticides (total)	0.005		
Polycyclic aromatic hydrocarbons (PAH)	0.2 ppb (0.002 ppm)		
Surfactants	200		
Radioactivity			
Gross beta	1000 pc/L		
Radium-226	3 pc/L		
Strontium-90	10 pc/L		
Bacteriological parameters			
Coliform cells/100 mL	100		
Total bacteria count/100 mL	1×10^6		<5000

Recreation and Aesthetics

Surface waters used for recreation and aesthetics should be free from :

(a) Substances which give colour, taste or turbidity (*e.g.* oils, grease, phenols, etc.);

(b) Materials which may yield objectionable deposits or float on the surface as bebris, oil and scum;

(c) Toxic substances which also include radionuclides, physiologically harmful to man, fish or other aquatic plants or animals; and

(d) Substances which may be promoting the growth of undesirable aquatic life.

Aquatic Life, Fish and Wild Life

In general, surface water has to be free from :

(a) Settleable matter which is able to inhibit the growth of flora and biota;

(b) Floating matter, like oil and grease, which tends to yield visible colour films on the surface, coats banks and bottom of water courses taints aquatic biota and brings about toxicity to fish and man. The temperature should not increase by 3 – 5 °F by hot waste water while pH and D.O. is within the limits prescribed in Table 1.1.

Irrigation Water

Salinity or total dissolved solids has been considered to be the most important parameter for irrigation water because it has the capacity to control the availability of water to plants through osmotic pressure-regulating mechanisms. The tolerance limits for trace elements have been given (in ppm).

Al 1.0; As 1.0; Be, 0.5; B 0.75; Cd 0.005; Cr 5.0;

Co 0.2; Cu 0.2; Pb 5.0; Li 5.0; Mn 2 0; Mo 0.005;

Ni 0.5; Se 0.05; V 10.0, Zn 5.0.

Some typical data for surface water have been given in Table 1.2. In general, ground water (well and tube well water) are having higher dissolved-mineral contents than surface waters. It is due to the intimate contact between CO_2 bearing water and rocks and soils in the ground and the length of time needed for dissolution.

TABLE 1.2

Typical Analysis of Some Surface and Ground Waters

Constituent, ppm	*Water from San Francisco Region*	*Water from Niagara Falls*	*Well Water of Ohio*
Silica	9.5	1.2	10
Iron (III)	0.07	0.02	0.09
Calcium	4.0	36	92

Magnesium	1.1	8.1	34
Total hardness (as $CaCO_3$)	14.6	123	169
Sodium	2.6	6.5	8.2
Potassium	0.6	1.2	1.4
Bicarbonate	18.3	119	33)
Sulphate	1.6	22	84
Chloride	2.0	13	9.6
Nitrate	0.41	0.1	13
Total dissolved solid	34	165	4'4

Sources of Water Pollution

Water pollution can result from natural runoff; dissolved chemicals in water that percolates through the soil; and human sources, such as agriculture, mining, construction, industry, homes, and businesses. It is useful to classify pollution sources as either point sources or non point sources. Point sources involve discharge of wastes from identifiable points. They include :

1. Sewage treatment plants (which remove some but not all pollutants)
2. Storm water runoff from combined storm and sanitary sewer lines in urban areas;
3. Industrial plants; and
4. Animal feedlots where large numbers (sometimes 10,000 or more) of animals are scientifically fed in tightly restricted quarters before slaughter. Non-point sources involve the diffuse discharge of wastes from land runoff, atmospheric washout, and sources that are difficult to identify and control.

They include :

1. Runoff of sediment from natural and human caused forest fires, construction, logging, and farming ;
2. Runoff of chemical fertilizers, pesticides, and saline irrigation water from croplands;

3. Urban storm water runoff;
4. Drainage of acids, minerals, and sediments from active and abandoned mines; and
5. Spills of oil and other hazardous materials. Non-point source pollution is not recognized as a major problem, since the sources are widely spread out, difficult to identify, and hard to control.

Water Pollution by Agriculture

Agriculture has been a victim of water pollution in many instances, but sometimes. It is also responsible for polluting water. Water pollution caused by agriculture is mainly an outcome of fertilizers and agricultural chemicals such as insecticides and herbicides.

1. *Water Pollution by Fertilizers.* Fertilizers given to crops are not always fully consumed, and part of it remains in the soil by being absorbed by soil colloid, and influence the quality of underground, river and sea waters when it is dissolved. Chemical fertilizers mainly consists of relatively simple compounds of nitrogen, phosphorus and potassium which are the nutritive elements of plants. Their runoff coefficients vary depending on the solubility of the fertilizer itself, the rate of absorption by plants, rate of absorption by the soil and rate of decomposition. It is generally believed that the runoff co-efficient of nitrogen is 5.30%, and that of phosphorus is 0.5-5%. Especially, nitrogen having a high runoff co-efficient tends to cause the eutrophication of lakes and enclosed sea areas.

   Fertilizers as the sources of water pollution are called "non-point sources" in contrast to "point-sources" such as sewage and industrial waste waters. The non-point sources are not only hard to eliminate, but also hard to measure their quantity. Organic fertilizers such as compost, farmyard manure, human waste and fish meal produce nitrogen and phosphorus when they are decomposed, and also become the sources of pollution.

These organic fertilizers do not contain, in general, any toxic substances.

2. *Water Pollution by Agricultural Chemicals.* Agricultural chemicals comprise are considered as a source of peculiar pollution which cannot be overlooked. Agricultural chemicals comprise a variety of chemicals such as insecticides, sterilizers and weed killers, and have played important roles in the operation of modern agriculture, but they have also caused serious environmental pollution.

The characteristics of agricultural chemicals as water-polluting substances lie in that : (1) almost all of them are special compounds not occurring at all in nature, and many of the behaviours and changes of such chemicals at the time when they are discharged into the natural environment are unknown; (2) many of their influences on the human body because of their inherent characteristic; (3) when they are ingested into the human body, they tend to accumulate or concentrate in it; (4) they are available in a variety of kinds, and new products are developed constantly; (5) some of them are extremely stable as chemicals, and remain in nature or in the human body for a long time thereby eventually afflicting the human body; (6) it is common to all such chemicals to contain additives besides the main constituents, and the additives can contribute to affect the environment though they do not have a strong toxicity; (7) many of them require a special method of analysis.

Of the agricultural chemicals developed in the past, many of them are prohibited from being used because of their violent toxicity. Nevertheless, those remaining in the environment are still factors that contribute to environmental pollution.

In addition to above-mentioned pollutions, there are instances of water pollution caused by indirect effect of agricultural activities. For example, the change in the condition of the ground surface by reclamation of waste land, disafforestation, supply of the soil to low land, and the alteration of the natural water system for irrigation purposes can be considered indirect causes of water pollution.

Water Pollution by Mining Industry

In many countries, water pollution related to the mining industry can be traced back to olden times compared with those caused by other industries.

The characteristics of waste water from mining industries varies greatly depending on the kind of mine. For instance :

(*i*) *Metal Mining Industry.* Many kinds of waste waters are discharged. For example, mine water, ore dressing water, waste water from smelters, etc. In general, it is strongly acid (low pH), and contains metal irons, sulphuric acid and suspended solids of high concentrations.

(*ii*) *Coal Mining Industry.* Waste water from coal mines mainly consists of mine water, coal dressing water, exudative water from coal-waste heap etc., and contains a large quantity of suspended solids. In general, the scales of operation of coal mines in much larger than that of other mining industries, and thus the coal mining industry exerts a large influence upon the natural environment.

(*iii*) *Sulfur Mine.* In general, waste water from sulfur mines contains large quantities of sulfuric acid and iron and has a low pH value. Almost all sulfur mines are deserted; so the administration of deserted mines becomes a social and environmental problem.

(*iv*) *Petroleum and Natural Gas Drilling Industries.* Different from other mining industries, these industries are very often located in plains. Waste water generally consists of welled up mine water and waste water incidental to the drilling operations: It contains large quantities of salt and organic matters such as mineral oil.

(*v*) *Stone-quarying and Clay-supply Industries.* These industries supply lime stones, building stones, porcelain clay etc. Waste water from the sites of these industries generally contains large quantities of suspended solids.

The characteristics of industrial pollution common to the above mentioned mining industries can be summarized as follows :

(i) In many instances, the development of a mine itself changes its natural environment artificially, and such a change first affects the quality of water of the environment even if any kind of waste water is not discharged from mining activities.

(ii) In areas where mines can be developed, very often, even before the development of the mine, the quality of river water of underground water is different from that of other areas. For instance, river water or underground water in such areas contain special metals of high concentrations and is strongly acid. Thus, it is rather difficult to measure the direct influence of mining activities on the quality of water. And also, geological features and the ways of land use are usually very peculiar in mining areas. These conditions should be taken into consideration prior to surveying the quality of water.

(iii) Mining activities generate large quantities of solid wastes in most cases, and such wastes from large heaps. Such wastes not only pollutes water by exudation of toxic substances, but also, the rapid generation of suspended solids in water when the heap is washed away by a heavy rainfall or a flood.

Water Pollution by Stockbreeding and Fisheries

1. When cattle or horses are fed on a big pasture, there is little possibility of the livestock causing water pollution because their excreta are spread widely all over the pasture. However, when livestock is fed intensively in a limited area it causes a number of problems. In the case of the stockbreeding industry. the excreta and left-over feed on livestock are the sources of water pollution. Water pollution caused by stockbreeding has the following characteristics.

(i) Degree of pollution varies largely depending on the location of the stockbreeding farm (whether in suburban or country area), system and scale of operation and the level of breeding technique.

(ii) Composition and quantity of excreta of livestock varies depending on the type, quantity and methods of feeding and watering.

(iii) Besides water pollution, the stockbreeding industry also generates an offensive odor, flies, noise and so on.

2. Besides stockbreeding farms, waste water from slaughter-houses is a source of water pollution which should not be overlooked.

3. In most instances, fisheries has been the victim of water pollution. But in recent years, as the number of fish farms increase, pollution of water areas surrounding fish farms caused by leftover feed and excreta of fish is becoming serious.

Water Pollution by Urban Activity

In a sense, the existence of human beings itself is a source of water pollution. In primitive ages, the purifying capacity of nature was much greater than the rate of water pollution caused by human living. But as time passed, men began to settle near water areas for the convenience of living; then they developed various industries which provided them with the foundation of civilized life. The emergence of various industries coupled with the development of civilization has constantly increased the quantity of waste water containing pollutants. On the other hand, the increasing concentration of the population has also invited the concentration of pollution sources. As a result, the rate of environmental pollution has exceeded the rate of natural purification.

If all the urban activities of human beings related with water pollution are viewed as a metabolic system, it can be outlined as follows :

		Waste water
Water		Human waste and municipal waste
Substances	Urban community	
Energies		Waste gas and waste heat

Municipal wastes such as waste water, human waste, household waste, waste gas and waste heat are the causes of water pollution.

The type of water pollu ion in a city varies largely depending on the cha'acteristics of the city, life style of the inhabitants, degree of civilization, degree of development of sewage treatment plants or sewerage systems. Although, it is very difficult to investigate these conditions in detail, they can be estimated by investigating the following :

(i) Population of the city : This is the most essential datum to be obtained first.

(ii) Consumption of water and quantity of waste water; where a city water system is wellestablished, it is relatively easy to obtain data concerning the consumption of water. For example, the consumption of city water in a modern civilized city is said to be about 500 liter/man/day. However, it is usually difficult to know the quantity of waste water even if it is only an estimate.

(iii) Condition of sewage treatme .t : This can be measured in terms of the availability of ·age treatment plants and sewage works including their capacities and efficiency.

(iv) Condition of municipal or public facilities and private facilities : In a large city, there are a number of public facilities and various other facilities, and all these facilities can be the sources of water pollution, such as : (Public facilities) Government and municipal offices, schools, hospitals, public bath houses, public lavatories, markets, ports, power plants, city gas facilities, sewage treatment plants, etc.

(Non-public facilities) Factories, stores, offices, gas stations, restaurants, bars, hotels, etc.

(v) Condition of sewage : Municipal sewage mainly consists of a mixture of human waste, household sewage and wastes from other facilities, and its condition varies depending on the characteristics of the city, degree of civilization, precipitation and the condition of road surface.

The quantity and condition of waste water resulting from human activity, mainly consisting of household waste water, generally vary depending on the time in a day; however, the "product of the concentration multiplied by the quantity" tends to remain within a relatively fixed range. Such a value is defined as "pollution loading amount", and is used as the basic data for the sewage treatment planning.

TABLE 1.3

Standard values of BOD, SS and Pollutant Load of Sewage and Human Waste

Type of Waste Water	*Load, Equivalent, Concentration, etc. of BOD and SS*
1	2
Human waste from flush toilets (including those from septic tanks for sewage)	1. BOD equivalent per person per day : 13 g (BOD : 260 mg/l) 2. SS equivalent per person per day; 26 g (SS : 520 mg/l) 3. Flushing water per person per day : 50l.
Household waste water (not including rainwater and special waste water)	1. BOD equivalent per person per day : 40 g (BOD : 200 mg/l)

(Contd.)

1	2
	2. SS equivalent per person per day : 50 g (SS : 260 mg/l)
	3. Quantity of waste water per person per day : 200 l
Other typical waste waters	1. Average BOD of waste water from office building; about 80 mg/l
	2. Average BOD of waste water from hotels; about 100 mg/l
	3. Average BOD of waste water from hospitals about 100 mg/l
	4. Average BOD of waste water from restaurants and catering facilities : about 400 mg/l

The most conspicuous characteristics of municipal sewage are that the organic content (indicates as BOD) is extremely high; the contents of ammoniacal and organic nitrogen, phosphates and sulfides are considerably high; and normally, hazardous substances* are not contained.

The modern sewage treatment technique is capable of removing organic pollutants very efficiently, but it is difficult economically to remove nitrogen and phosphorus contents efficiently with existing facilities. For example, only about 30%

* Industrial waste water tends to contain hazardous substances; therefore, its discharge into the municipal sewerage system should be strictly controlled. This is necessary, because ordinary sewage treatment plants not only do not have an equipment to remove hazardous substances, but also there is a possibility of getting its activated sludge processing function damaged by such hazardous substances.

of the nitrogen and phosphorus contents can be removed through the currently employed two-step treatment method, which mainly applies the activated sludge method, and the remainder of such contents causes the eutrophication of water areas.

Anyhow, all of the above mentioned techniques are possible where the municipal sewerage system and sewage treatment plants are well established, and the quantity and quality of waste water can be measured effectively. However, where sewage treatment plants are not available, or where waste water is discharged directly into the hydrosphere without passing a sewage treatment plant, it is almost impossible to measure the quality and quantity of waste water.

In case waste water is directly discharged into the hydrosphere, and if the pathogenic bacteria of an infectious disease is contained in the human waste, such bacteria would be discharged into the hydrosphere in active condition thereby eventually spreading the infections disease through the water system. The degree of such a danger various depending on the quality of sanitation facilities of individual cities.

WATER POLLUTION BY MANUFACTURING INDUSTRY

Characteristics of Water Pollution by Manufacturing Industry

Of water pollutions with which human activities are concerned, the manufacturing industry is the largest in scale, and most diversified. The characteristics of the water pollution of manufacturing industry can be summarized as follows :

1. Unlike the mining industry and agriculture, the discharging condition of waste water from manufacturing industries is relatively simple because it is rarely affected by complicated conditions of the natural environment. For example, waste water of the manufacturing industry is discharged concentratively from a fixed outlet, which is called "point source pollution". Besides, waste water from the manufac-

turing industry is generated as a result of the production activity which is 00% artificial, and thus the manufacturing industry can be considered as the source of water pollution which is most easy to catch and control in this point.

2. Sources of water pollution from manufacturing industry are relatively easy to trace back through scientific method, but their control is not always easy because it involves complex social and political problems. Furthermore, tracing such point sources of pollution is not always easy because the interests of individual manufacturing companies can be affected by pollution control measures.

3. In contrast to the relatively simple form of discharging process, the composition and quantity of waste water vary widely depending on the type and scale of individual factories, change from time to time, and its influence also takes a variety of forms.

4. The influence of waste water on the environment varies depending on the geographical location of the factory. For example,

   (i) In the case of an industrial conglomerate or large scale industrial complex : the influence of waste water on the environment is extremely large, but as far as such manufacturing facilities are concentrated in one area based on the results of careful environmental assessment, waste water from these facilities can be controlled in the most ideal form;

   (ii) In case factories are located in commercial or residential areas of a city : the location of this type is primarily undesirable. Actually, however, there are a number of such cases because of historical reasons, or geographical advantages concerning the purchase of raw materials, transportation of products, or access to the labour market. In Japan, locations within easy access to the source of industrial water are decreasing in these areas. As a result, factories that recycle water are increasing, small scale manufacturing companies establish cooperative associations and are promoting the construction of joint waste water treatment facilities.

(iii) Location of the manufacturing factory is whether mountainous area, plain, or costal region is also an important factor to be considered is assessing the influence of waste water on the environment. Especially, in the case of a water-consuming type of manufacturing industry, its location has to be decided depending on accessibility to an industrial water source. On the other hand, the condition of the river or the sea into which waste water of the factory has to be discharged also has a great influence on the form and type of environmental pollution.

Characteristics of Waste Water of Main Industries

As discussed in the forgoing, the composition of industrial waste water is complex, and it varies largely according to the difference in the manufacturing process even among the same type of manufacturing factories. Furthermore, the manufacturing industry is constantly subject to the impact of technological innovations which occur at increasingly short intervals. For example, in Japan, the technique for recycling of waste water is making rapid progress. Manufacturing companies were compelled to do so due to the increasingly strict pollution control requirements and the increasing shortage of industrial water. As a result, manufacturing companies are also compelled to alter conventional manufacturing processes. It is rather difficult to enumerate the characteristics of waste water of each industry, General characteristics of waste water according to the type of manufacturing industry are as follows :

1. *Food Processing Industry.* A common characteristic of waste water from all types of food processing industries is the high concentration of organic content. Another

common characteristic is the high concentration of solid, semisolid and suspended solid contents in waste water discharged from raw material treatment processes. The quality and quantity of waste water vary depending largely on the season mainly due to the variation of raw materials to be processed. Materials contained in waste water can be divided roughly into carbohydrate, protein, fat and a mixture of them. In many instances, waste water contains oil, nitrogen and phosphorus. In the case of starch producing, sugar refining and brewing industries, the BOD level of waste water is especially high.

2. *Textile Industry*. There is much difference in the content of waste water between natural and chemical textile industries. Waste water from the wool industry contains high concentrations of BOD, fat and alkali. Waste water from dyeing discharges various pollutants such as dyes; auxiliary additives and other chemicals.

3. *Paper and Pulp Making Industries*. These are typical water-consuming industries and the levels of COD and suspended solids in waste water are extremely high. Especially, waste water from the digesting process contains all components of wood except cellulose and has high COD and dense color, so that it is necessary to collect and treat most of the content to recover. A large quantity of waste paper fibre is also discharged as suspended solids together with the waste water.

4. *Petroleum Refining Industry*. The main component of pollutants is spilled oil; but waste water also contains hydrogen sulfide, other sulfides, ammonia, mercaptan, and phenols, etc.

5. *Petrochemical Industry*. The waste water contains hydrocarbons, various organic compounds and catalysts, and its content varies largely depending on the type of product.

6. *Iron and Steel Making Industry*. This industry is also a water-consuming industry. Waste water from the

cooling and cleaning processes for coke furnace gas contains ammonia, cyanide, phenols, etc. Besides, there is waste water from the dust collecting process of each furnace which contains suspended solids (coke dust and ore), and that from the pickling process which contains acid, iron and oil.

7. *Electroplating Industry*. Many of the raw materials and chemicals used in this industry contain toxic substances, and thus the waste water contains various heavy metals (cadmium, zinc, copper etc), cyanide, hexavalent chromium, acids and alkalis.
8. *Leather Industry*. The level of BOD is high in waste water which comes from tannin used for tanning and raw skins. It also contains chromium used for tanning and lime for pretreatment, suspended solids and dyes.
9. *Steam and Nuclear Power Plants*. These power plants discharge large quantities of effluent water from the cooling process. In Japan, all the large power plants are located close to the sea, and these plants are using sea water for cooling. The temperature of waste cooling water is normally high, and such waste water can cause thermal pollution or the elevation of temperature to the vicinity of discharging outlet. No other pollution is considered.

Water Pollution by Other Human Activities

Aside from the various human activities already discussed in previous sections, there are other types of human activities which also cause water pollution, and such activities sometimes become social problems as sources of environmental pollution. These types of human activities are as follows :

1. *Construction Work*. Construction works, especially those large-scale land development projects or forestry conservation and flood control projects aimed at preventing natural disasters, sometimes can cause water pollution. Water pollution caused by such construction

projects can be divided into those occuring incidental to the progress of construction works and those occurring due to the change in natural conditions as a result of the construction works. The latter is more important than the former. Construction works for a dam, river improvement (including those for the alteration of basins), port facilities, creation and reclamation of land and gravel gathering from river-bed are enumerated as examples of the above-mentioned cases.

2. *Various Tertiary Industries.* Various tertiary industries can be considered to cause water pollution such as cleaning and laundry businesses, public bath houses, hotels, printing business, photographic film development businesses, hospitals, laboratories or research institutes, tourist businesses, transportation facilities and so forth. An increase in the load of municipal waste water as a result of the increase in the load of municipal waste water as a result of the increase in the working population of cities also has become one of the factors directly related to water pollution. Of various transportations, ships, especially tankers which discharge waste cleaning water, have come to the fore as sources of water pollution in coastal regions.

3. *Errors, Accidents and Disasters.* Accidental damages to public sewage works or drainage of industrial effluent, as well as poor planning of sewage treatment plants, and erroneous operations of these facilities can also cause water pollution. Speaking of water pollution caused by accidents, pollution of the ocean due to oil resulting from the collision of tankers on the high seas has become an object of public attention in recent years.

Eutrophication

Phosphorous and nitrogen compounds (phosphates, ammonium salts, nitrites and nitrates, etc.) are called nutrient salts which are essential to the growth of phytoplanktons and algae participating in the primary production. However, if they are excessively supplied together with organic substances, the production activities become too vigorous due to overnutrition, ultimately changing the aquatic ecosystem. This phenomenon is called eutrophication. Eutrophication of lakes is not an extraordinary phenomenon but one which occurs and slowly proceeds even in natural not polluted ecosystem. However, eutrophication in polluted sea and lakes progresses very quickly due to rapid increase of organic substances and nutrient salts, causing unusual growth of specific organisms and thus destroying the normal aquiatic ecosystem. This type of eutrophication is often observed in closed water areas, such as lakes and deeply recessed bays; the results is coloring and smelling of the drinking water taken from the lakes, lowered clarity of lake water, or the occurrence of the so called "red tide" in sea and lakes as a visible and conspicuous phenomenon. The red tide is caused by unusual propagation of certain kinds of planktons at such a terrific speed as to change the color of water.

Consequently, many environmental trouble occur such as death of fishes due to clogging of gills with planktons, oxygen deficiency over the polluted area, and generation of poisonous substances.

In Japan, this phenomenon has often been observed in the Inland Sea and other closed water areas. The true causes have not yet been determined exactly, but it is generally accepted that phosphorus and nitrogen compounds contained in urban and industrial waste waters and playing an important role.

Contamination of Bottom Sediments

River, lake and sea bottoms are usually covered with a certain depth of muddy sediments. These sediments contain remains and excreta of aquatic organisms, other than the earth and sand carried over by rivers. Suspended solids in water tend to sediment remarkably in lakes, slow-flowing rivers, estuaries where the river stream is stagnated by tidal currents, and in closed sea areas.

Cases are also reported where some highly poisonous heavy metals and sparingly soluble organic substances, which had settled on the bottom of water and fixed as sediment over a long period, were dissolved again or raised under some conditions, causing serious hazards. A typical example is mercúry in the bottom sediment of the Minamata Bay. It was about twenty years ago when mercury caused the Minamata disease, but mercury hazard are still continuing Minamata disease, but mercury hazards are still continuing even after the discharge of mercury from industrial waste water completely stopped.

At the bottoms of lakes and closed sea areas with high organic loads, dissolved oxygen is consumed rapidly during the putrefaction of organic substances contained in the bottom mud. If vertical mixing of water is insufficient due to stratification, oxygen dissolved on the surface of water from the atmospheric air cannot reach the bottom. In addition, poor clarity of water weakens sufficient penetration of the sun beam and significantly suppresses photosynthetic reactions in the bottom water layer. Under these conditions, the dissolved oxygen in the bottom water mass will decrease, ultimately leading to the zero-oxygen state. As a result, fishes in that area will be seriously affected and the surrounding ecosystem may be materially destroyed.*

Other cases are also known where elution of phosphorus and nitrogen from bottom sediments in closed sea areas may be responsible for eutrophication and the resultant occurrence of red tide. Water bottoms often provide favourable spawning grounds for fishes; anaerobic conditions of bottom sediments destroy hatching in many cases.

---

* The same situation is also observed in nature, but not as a result of artificial pollutions. A typical example is observed in the Black Sea. The seawater flowing from the Mediterranean Sea through the strait settles on the bottom, which substantially shuts off the vertical circulation of water. Consequently, putrefaction of sedimented organic substances proceeds extensively, generating a large amount of hydrogen sulfide. The botrom water layer has no dissolved oxygen, and no fishes exist there.

A survey of bottom sediments provides the environmental information of the neighboring water area in the past cver a considerably long pcriod. Thus, it is possible to obtain new findings on the progress of pollution not expected from water quality investigations.

Oil Pollution in Oceans

Modern civilization relies in a large measure on petroleum in all aspects. A huge amount of crude oil is carried by super-tankers on oceans throughout the world, stored in tanks at terminal stations, and processed into various petroleum products at petrochemical complexes. Many terminal stations and petrochemical complexes are located in coastal zones. Probably, a considerable amount of oils is always flowing into the ocean, including seight leaks from various ground facilities, leakage from tankers during loading and unloading, and discharge of waste oil used on land. Besides, large scale oil spills due to maritime accidents have often occured.

Oil spilt out to the sea tends to-spread on the sea surface as a thin layer. Volatile components gradually evaporate and some of them dissolve into water (benzene, for example, dissolves in seawater to a concentration of about 820 ppm at ordinary temperature). The oil membrane thus formed on the sea surface disturbs normal exchange of gases between the seawater and the atmospheric air, and also interrupts the penetration of sun beams to retard the growth of planktons.

The oil gradually becomes thick as the volatile components evaporate; the resultant viscous heavy oil is in or on the sea-water, or gathers shellfishes and sand on the sea bottom to form oil balls, which drift over long distances and cast ashore. The spilt oil is degraded by some bacteria through very slowly. However, if it drifts faster than it decomposes, pollution will proceed. Heavy oil sometimes sticks to the bodles of seafowls and fishes; particularly, larvas are seriously damaged by oil deposition,

Although not so critical, cases are often encountered when fishes cannot be used for food because of the smell of heavy oil.

Dissolved Oxygen

Oxygen dissolved in water has been an important vital species which gets consumed by oxidation of organic matter/ reducing agent etc. It is regarded as an important water quality parameter. Its optimum value for good quality water has been 4·6 mg/litre of dissolved oxygen (DO) which is able to maintain aquatic life in a water body. If DO values are somewhat lower than this value, this indicates water pollution.

Biochemical Oxygen Demand (BOD)

It is a water quality parameter for organic matter in water, which is empirical in nature. It may be measured by the quantity of oxygen utilised by suitable aquatic micro-organisms during a five day period.

Biochemical oxygen demand (BOD) has been a standardized measurement of the amount of oxygen that would be required by micro-organisms to cause the decomposition of certain organic and inorganic matter in the water. The measurement is done under standardized conditions (*e.g.*, at 20°C and five days to allow the decomposition to take place). The result is called the 5-day BOD and is expressed in milligrams of oxygen per litre of water. BOD is not a pollutant but an indicator. It measures no particular substance but a family - any substance that micro-organisms can consume (using oxygen as they do) or any material attacked under the conditions of the test. The substances decomposed in the test may be food used by the micro-organisms or certain chemicals that are readily attacked by oxygen, perhaps with the aid of enzymes released by the micro-organisms. These chemicals include sulphites and sulphides (from paper mills), ferrous iron, and some easily oxidized compounds. Many organics, however, make no contribution to the BOD but still render the water unfit for human use. BOD values of several hundred milligrams per litre characterize "strong" sewage. For "excellent" drinking water the 5 day BOD, on a monthly average, should be in the range of 0.75 – 1.5 milligrams/litre. BOD values are important when they signify that the oxygen supply dissolved in the water will be so greatly reduced that desirable fish no longer can survive or when they signify that conditions for the propagation of dangerous bacteria exist.

Chemical Oxygen Demand

It is an important water quality parameter. It is an index of the organic content of water oxygen demanding substances in water.

The chemical oxygen demand (COD) has been a measure of the concentration in a water supply of substances that can get attacked by a strong chemical oxidizing agent in a standardized analysis. (Dichromate oxidation is commonly used). The results of the analysis are usually expressed in terms of the amount of oxygen that would be needed (in principle, because oxygen is not itself used) to oxidize the contaminants to the same final products obtained with the standardized analysis. COD values do not get necessarily correlated with BOD values. Textile wastes, paper mill wastes, and other wastes with high levels of cellulose have COD values considerably higher than their BOD values as cellulose is not readily attacked in the BOD test. Distillery and refinery wastes often have BODs higher than CODs unless the COD measurement is specialy modified. In the nature of the two tests, the BOD of a given water supply is able to decrease faster than its COD.

Chemical oxygen demand (COD) has been found to be more scientific than the traditional empirical concept of Biological oxygen demand (BOD). The test is based on the chemical oxidation of material in water by $K_2Cr_2O_7$ in H_2SO_4.

A comparative account of BOD and COD is given in Table.

TABLE

Comparison of BOD and COD

Biological Oxygen Demand (BOD)	*Chemical Oxygen Demand (COD)*
(i) It may be defined as the amount of oxygen used for biochemical oxidation by microorganisms in a unit volume of water. This test has been developed for five days at 20°C.	(i) It may be defined as the amount of oxygen required by organic matter in a sample of water for its oxidation by a strong chemical oxidant, and is expressed as ppm of oxygen taken from a solution of potassium dichromate in two hours.
(ii) As BOD value approximates the amount of oxidizable organic matter, it is therefore, used as a measure of degree of water pollution and waste strength.	(ii) This value has been a poor measure of strength of organic matter as oxygen also get consumed in the oxidation of inorganic matter such as nitrates, sulphates, reduced metal ions, and also that some organic materials like benzene, pyridine and few other cyclic organic compounds do not get oxidized by this test.
(iii) BOD values have been useful in process design	(iii) It has been a very important parameter in manage-

and loading calculations, measure of treatment efficiency and operation, stream pollution control and in evaluating self purification capacity of a stream.	ment and design of the treatment plants due to its rapidity in determination. Values are taken as basis for calculation of efficiency of treatment plants and also figure in the standards for discharging industrial domestic effluents in various kinds of waters.
(iv) Types of microorganisms, pH presence of toxins, some reduced mineral matter and nitrification process have been the important factors that influence the BOD test.	(iv) Presence of toxins and other such unfavourable conditions for the growth of microorganisms are not able to effect COD values.

2

Hydrosphere

Introduction

Water is a very complicated and vitally important substance. It is the medium which gave birth to the first primitive living molecules, and without it no life can exist.

Throughout history, the quality and quantity of water available to man have been vital factors in ascertaining his well-being. Whole civilizations have disappeared due to water shortages resulting from changes in climate, and the process continues today with the starvation of African peoples as the Sahara marches southward. Disastrous floods have always plagued mankind. Waterborne diseases are able to kill millions of people although in the more well-developed areas of the world, the great plagues of cholera and typhoid have been controlled. Ambitious programs of dam and like construction have reduced flood damage, though not without undesirable side effect. However, problems with water remain, and in some respects have been becoming more serious.

Any consideration of environmental chemistry as applied to water needs some understanding of the sources, transport, characteristics, and composition of water. The chemical reactions taking place in water, and the chemical species found in it, get influenced strongly by the environment in which the water is found. The chemistry of water exposed to the atmosphere has been quite different from that of water at the bottom of a lake. Microorganisms play a vital role in determining the chemical composition of water. Thus in discussing water chemistry, one must discuss the many general factors which are able to influence this chemistry.

Hydrological Cycle

The hydrologic cycle refers to the process whereby water gets converted from its liquid or solid state into its vapor state. As a vapor the water has been capable of travelling considerable distances from its source prior to recondensing and returning to earth as precipitation. Hence the hydrologic cycle has been a complex, interrelated system which is involving the movement of atmospheric, surface (marine and fresh), and groundwater throughout various regions of the world. It has been the hydrologic cycle which has been solely responsible for the world's precipitation, and it is this precipitation, falling on the terrestrial and surface freshwater environments, that has been the sole source of the earth's supply of fresh water (Fig. 3.1).

The hydrologic cycle may be having either a long or various short cycles. In the short cycles, water may evaporate from either marine or freshwater systems, condense almost immediately, and return as precipitation to the same system. Another variation of a short cycle has been the precipitation and subsequent evaporation of water from land surfaces, which has been followed by its condensation and return as precipitation to the land, followed by reevaporation, and so on.

In the long cycle the major source of water vapor has been the world's oceans, which are having 97.3% of the earth's waters. In this cycle a portion of the water evaporates and forms clouds that move inland. The water vapor then cools and returns to earth as precipitation. It is estimated that only 0.007% of the oceanic water has been distributed to terrestrial areas annually. This water will ultimately return to the oceans through river and groundwater flow. As precipitation may take place close to the source of initial evaporation or thousands of miles away, the water may remain in the vapor state for

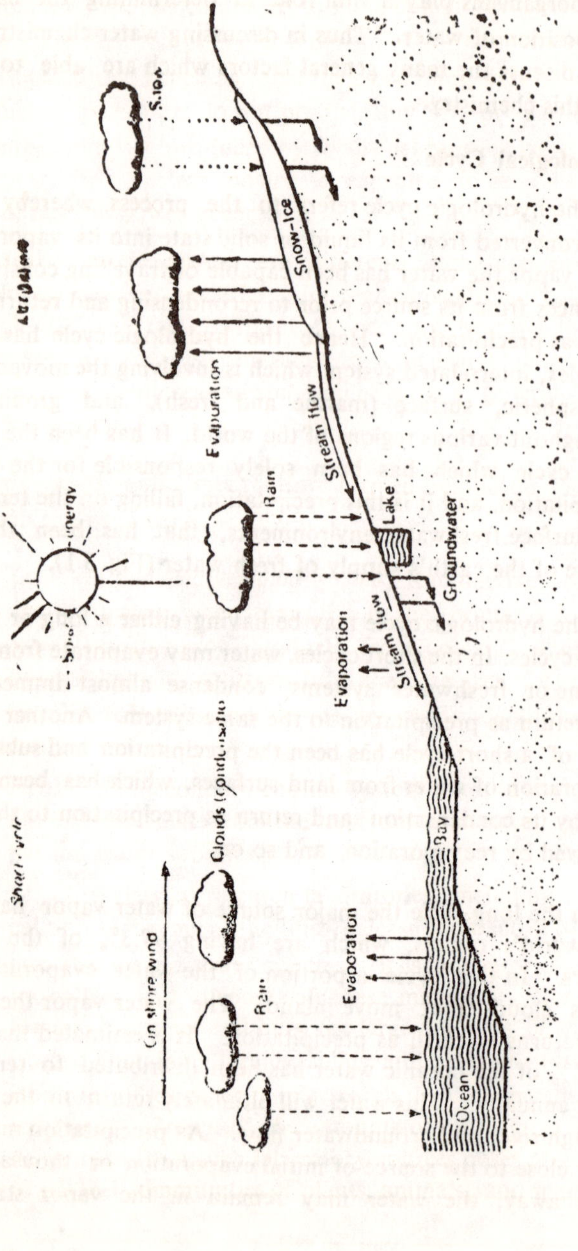

Fig. 2.1 The Hydrologic Cycle

variable times (a few hours to a few weeks). The average residence time for water to remain in the atmosphere is regarded to be 10 days.

The various stages in the hydrologic cycle are as follows ;

1. Evaporation and Condensation

Water vapor gets evaporated primarily from the oceans and then tends to condense around minute particles, termed nuclei, which have been suspended in the atmosphere. The nuclei generally are having small particles of organic material (spores, pollen, etc.), fine mineral particles, volcanic ash, and the like. Dust and smoke particles from industrial sources and automotive exhausts may be serving as nuclei and form a major factor in contributing to the contamination of rainwater. This is due to the fact that airborne particles (such as lead) readily dissolve in the newly condensed atmospheric water, and these materials then gets returned to earth along with the rainwater.

Initially, the condensed liquid has been in the form of extremely small droplets [less than 0.04 millimetre (mm) in diameter]. Due to their 'small size, their rate of fall has been negligible, and they get retained in the atmosphere as clouds. Finally, however, this vapor forms precipitation in response to one of three major factors. The largest amount of precipitation would be formed when masses of warm, moist air move into regions of cold air. This causes a rapid condensation of the vapor and subsequent precipitation. The second factor arises during periods of warm weather, when air, warmed at the earth's surface, gets decreased in density and rises into overlying cold air, bringing about condensation of the water vapor. The third mechanism involved in condensation takes place with the cooling of air masses as they move over high mountains. In each case the cooling of warm, moist air has been responsible for the condensation and precipitation of rainwater.

2. Runoff, Stream Flow, and Infiltration

When rain is falling on a terrestrial area, a portion of the water would be caught by vegetation. This process is known as interception and this water readily get reevaporated, as there is a large surface area that is exposed to wind action. The remainder of the water falls to earth, and a portion sinks into the soil surface by a process known as infiltration. The portion not infiltrated into the soil, called surface runoff, flows over the surface and gets discharged (in undeveloped areas) into streams. The water entering a stream by surface runoff plus the water entering via groundwater flow is known as runoff. Thus the terms "runoff" and "surface runoff" have been different and distinct. Surface runoff equals precipitation minus the water lost by interception and infiltration. Runoff, on the other hand, has been generally synonymous with stream flow and has been equal to the sum of the surface water plus the groundwater that enters a stream.

Infiltration has been somewhat also distinct from groundwater when the water percolates into the earth's surface, different portions would be following three distinct pathways. A portion of the water will function as interflow, since the presence of impermeable lower sediments will disallow deep panetration of this water. Consequently, this segment will flow just below the soil surface and discharge into streams. Another portion would be remaining above the water table in an area called the zone of unsaturated flow. Both of these portions of water are regarded to be in the zone of aeration. The third portion will percolate down into the groundwater table and will, eventually, get discharged into streams.

3. Evapotranspiration

A large amount of the precipitated water would be converted to vapor by evaporation and/or transpiration. Evaporation refers to the process whereby molecules of liquid water (at the surface of a water body or in moist soil) absorb sufficient energy to leave the liquid state and enter the vapor state. The energy absorbed by the water in this process gets stored within the vapor - hence somewhat large amount of heat contained in the moist air.

Transpiration refers to the process whereby terrestrial and emergent aquatic vegetation release water vapor to the atmosphere. In the process of photosynthesis all plants have been taking in liquid water and carbon dioxide and, by a complex series of reactions, convert these materials to carbohydrate, oxygen (gas), and water vapor. Sub-mergent aquatic vegetation (plants growing completely beneath the water surface) also liberates water and oxygen as vapors but, in these cases, the oxygen and water vapour produced and released into the surrounding liquid water would immediately form hydrogen bonds with the liquid water. The water vapour will get converted to its liquid form, whereas the oxygen remains in the water column as dissolved oxygen. As the leaves of both terrestrial and aquatic emergent vegetation have been in air, the water vapour and oxygen, when released by these plants in photosynthesis, will remain in the vapour state as there are insufficient water molecules in the immediate vicinity to form hydrogen bonds and keep the water vapour and oxygen in solution.

The water converted from liquid to vapor in transpiration has been quite considerable, and the ability of plants to remove water from both soil and aquatic systems cannot be underestimated. Transpiration by emergent vegetation has been a major factor in the "drying up" of lakes.

Groundwater, unless it has been within a few feet of the surface, could not be evaporated. Rather, the portion that gets converted to vapour gets transpired by plants. In most regions the water lost by evaporation cannot be measured separately from the water lost transpiration. Consequently, the two are regarded together as evapotranspiration.

4. Ground Water

The most lengthy portion of the hydrologic cycle would get completed when groundwater gets returned to the earth's surface. The return may take place by springs, transpiration, or by artificial means.

Any natural surface discharge of sufficient water that will flow as a small rivulet is known as a spring, while a smaller discharge is known as surface seepage. Groundwater may also be discharged as subaqueous springs below the surface of lakes, rivers, and marine systems. It is possible to classify springs generally on the basis of their magnitude of discharge, as summarized in Table 2.1. Three major factors affecting the magnitude of a spring have been the land area accepting the rainfall and thus contributing water to the sub-surface system,

TABLE 2.1

Classification of Spring Discharge

Magnitude	*Discharge*
First	Greater than 2.83 m^3/s
Second	0.283-2.83 m^3/s
Third	2.83-283 litres/s
Fourth	2.83-6.31 litres/s
Fifth	6.31-0.631 litres/s
Sixth	63.1-631 ml/s
Seventh	7.9-63.1 ml/s
Eighth	Less than 7.9 ml/s

the permeability of the soil and subsoil, and the quantity of water entering the system (the amount of recharge). The use of water, in developed areas, prior to recharge has been known to alter the quality and quantity of the water entering the sub-surface system.

Discussion

The water used by man is primarily fresh surface water and ground water. In arid regions a small fraction of man's water supply comes from the ocean, a source which is likely to become more important as the world's supply of fresh water dwindles relative to demand. Saline or brackish ground waters may also be used in some areas.

Ground water and surface water are having appreciably different characteristics. Many substances either dissolve in surface water or get suspended in it on its path to the ocean. Surface water collected in a lake or reservoir, and having the mineral nutrients essential for algae growth, may support a heavy growth of algae. Surface water having a high level of biodegradable organic material normally contains a high population of bacteria. All these factors are having a profound effect upon the quality of surface water.

Ground water is able to dissolve minerals from the formations through which it passes. Most microorganisms originally present in ground water are slowly filtered out as it seeps through the ground. Occasionally, the content of undesirable salts may become exceseively high in ground water, although generally it has been superior to surface water for use by man as a domestic water source.

2.3 Unique Properties of Water

The study of water is termed as hydrology. This is divided into a number of subcategories

(*i*) *Limnology*. It is the branch of the science which deals with the characteristics of fresh water, including biological properties as well as chemical and physical properties.

(*ii*) *Oceanography*. It is the science of ocean and its properties.

Water is having a number of unique properties, without which life could not exist. Many of these properties have been due to hydrogen bonding in water. It has been an excellent solvent for many Materials. Thus water is the basic transport medium for the nutrients and waste products involved in life processes. The extremely high dielectric constant of water compared to other liquids is having a profound effect upon the solvent properties of water, in that most ionic materials are dissociated in water. With the exception of liquid ammonia, water is having the highest heat capacity of any liquid or solid, 1 cal × g^{-1} × deg^{-1}. Due to this high heat capacity, a relatively large amount of heat is needed to change the temperature of a mass of water appreciably, which can have a stabilizing effect upon the temperature of whole geographic regions. In addition, it does not allow sudden large changes of temperature in large bodies of water, and thereby protects aquatic organisms from the shock of abrupt temperature variations. The extremely high heat of vaporization of water, 585 cal/g at 20°C, likewise helps to stabilize the temperature of bodies of water and the surrounding geographic regions. It also influences the transfer of heat and water vapour between bodies of water and the atmosphere. Water is having its maximum density at a temperature (4°C) above its freezing point. The fortunate consequence of that fact has been that ice floats so that few large bodies of water ever freeze solid. Furthermore, the pattern of vertical circulation of water in lakes, a determining factor in their chemistry and biology, governed largely by the unique temperature-density relationship of water.

2.4. Biological Method of Zonation

As sunlight has been such an important factor in the productivity of these systems, it has been customary to divide both marine and freshwater systems into definite zones on the basis of the depth of sunlight penetration. This has been primarily a biological method of zonation. The various wavelengths of light are rapidly filtered out as they pass through a water column. Based upon the degree of light penetration, it has been possible to divide a system into certain zones (Fig. 2.2). The upper zone of sunlight penetration is known as the euphotic zone. Beneath this is a zone of perpetual darkness, the dysphotic zone, which is extending from the bottom of the euphotic zone to the bottom. The bottom sediment underlying the dysphotic zone is known as the benthic zone and the regeneration zone.

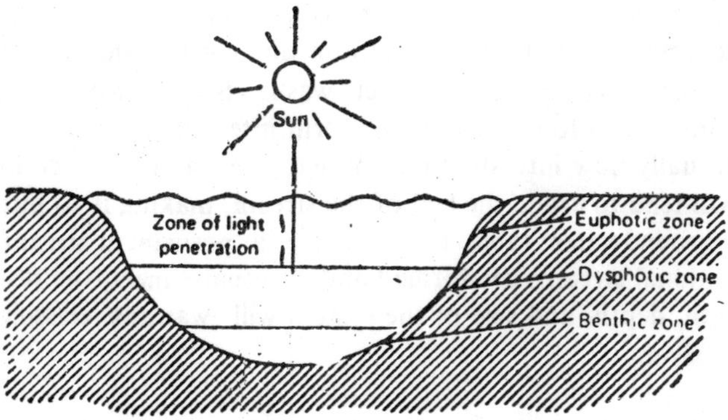

Fig. 2.2. Lake zonation on the basis of sunlight penetration.

According to this type of classification, shallow lakes, streams, and rivers would not behaving a dysphotic zone, as they would be shallow enough to allow for sunlight penetration down to the benthic zone. In addition, the position of both the euphotic and dysphotic zone in deep systems generally shifts with environmental conditions. For example, during periods of heavy rainfall, when sediment has been washed into the system, the physical presence of this suspended sediment would tend to block the sunlight, bringing about a reduction of the euphotic zone and an increase of the dysphotic zone.

In typical aquatic and marine systems the euphotic zone has been regarded to be both the zone of sunlight penetration and the zone of oxygen and food production. It has been in this zone that the plants and the primary consumers would be found, at least during the periods of active photosynthesis and grazing. The other consumers (secondary, tertiary, etc.) would occur in all three zones, depending upon the availability of their food source.

2.5. Lotic Environment

The lotic environment is consisting of all inland waters in which the entire water body continually moves in a definite direction. Thus rivers, streams, and brooks are regarded to be lotic environments. These systems have been generally originating from precipitation that falls on the earth's sloping surface and flows downhill as sheet wash. Sheet wash tends to accumulate in low areas and to form intermittent rivulets, which eventually flow into streams. When the water flow continues, it will tend to erode the bed of the stream, making it to intercept the groundwater table and become a permanent stream. As additional streams are formed and eventually meet, the resulting length, depth, and size of the system will warrant classification as a river.

The rivulet-stream sequence can still get observed upstream, but in an ever-changing position, because erosion will make an upstream migration of these features. The migration will continue until halted by a natural feature like a drainage divine. A direct result of this erosion-induced geological migration has been a continual migration of biological habitats. In other words, erosion will make the headwaters to "migrate" upstream which is followed by a similar "migration" of the environmental conditions characteristic of rivulets, streams, and rivers. The animal and plant communities that occupy each of these habitats must also move upstream, to remain in the same habitats (which has been in a constantly changing position because of erosion), or must adapt to the gradually changing conditions brought about by the erosion caused by the running waters. In natural, undisturbed systems these changes have been very slow and permit ample time for the organism to adapt to the changing conditions.

As lotic systems tend to be shallower and narrower than lakes, there has been generally a greater proportion of water exposed to land surfaces. Consequently, streams have been more intimately associated with and affected by the terrestrial environment than have been most lentic systems. In addition, lotic systems have been more dependent on the surrounding land areas for a large proportion of their nutrient supply.

2.6. Lentic Environment

The lentic environment has been including all inland waters in which the water has been not continually flowing in a definite direction. The water in these systems have been essentially standing, although some water motion may take place because of wind-driven waves and or in the vicinity of inlets and outlets. From a long-term geologic viewpoint, however, these inlets and outlets have been regarded to be temporary impoundments that will in time disappear, converting the system to a flowing-water environment. Typical lentic environments have been regarded to be lakes, ponds, and marshes.

In lentic systems the aging process has been just the reverse of that described for lotic systems. Whereas streams tend to get wider and deeper as they age, lakes tend to get shallower and the banks extend into what was originally open water. Natural filling has been generally due to wind-blown materials (sand, leaves, etc.) entering the system, sediment input by streams, and terrestrial runoff and aquatic plant and animal debris. Not all lentic environments have been becoming shallower by this type of filling alone, however. In many cases outlets may widen and deepen, making an increased out flow of water having a subsequent lowering of the water level.

Lakes could be generally classified, on the basis their depth and nutrient levels, as oligotrophic, eutrophic, or senescent.

An oligotrophic lake has been regarded to be in its young stage. These lakes have been very deep, with a sand or rock bottom, and have been low in nutrients. Because of the paucity of nutrients, both plants and animal life are low.

Eutrophic lakes have been regarded as middle-aged systems. They have been relatively shallow in comparison to oligotrophic systems, with a silty or mud bottom, and have sufficient nutrients for supporting a large population of plants and animals.

A senescent lake has been in the oldest stage of development. The bottom sediments in senescent system have been consisting of a thick layer of organic silts and/or muds, nutrient levels are high, and the system has been very shallow. There has been a large percentage of rooted emergent vegetation which is growing throughout the system. Terrestrial or marsh vegetation tends to grow along the banks and into the lake itself over the root mat.

On the basis of temperature (and thus, indirectly, density) lake could be divided into three distinct zones : the epilimnion, the thermocline, and the hypolimnion (Fig. 2.3). The epilimnion has been the upper zone of gradual temperature change. Below

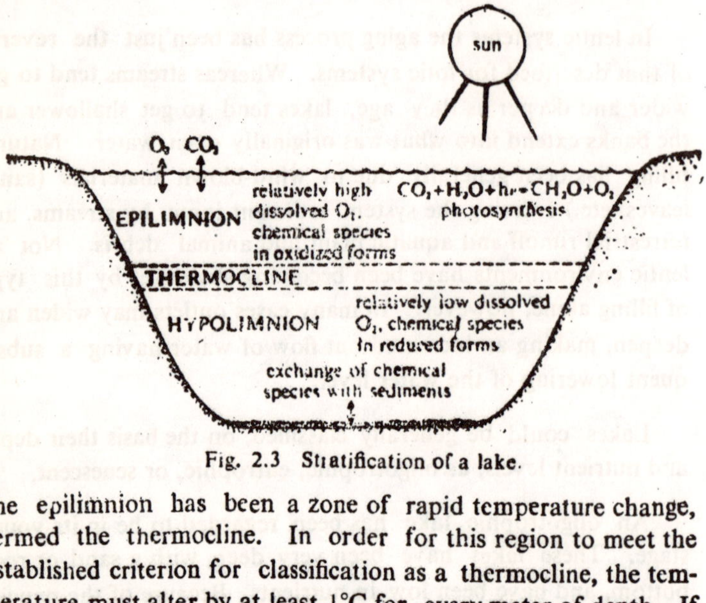

Fig. 2.3 Stratification of a lake.

the epilimnion has been a zone of rapid temperature change, termed the thermocline. In order for this region to meet the established criterion for classification as a thermocline, the temperature must alter by at least 1°C for every meter of depth. If the temperature changes at a lesser rate relative to depth, these water would be classified as a portion of the epilimnion. Below the thermocline there has been an area of water that has been constantly at a temperature of 4°C, termed the hypolimnion. As the waters of the hypolimnion have been at 4°C they have been at their maximum density and will occur at the bottom layer in any system. Comparing the relative water temperatures and density variations seasonally, it has been noted that in summer the warmer water will get located in the upper epilimnion and the colder, denser, water will be located in the hypolimnion. In the winter the colder; less dense water at temperatures below 4°C will be located in the hypolimnion. In either case, the less dense water has been in the epilimnion and the dense (4°C) water has been always found in the bottom waters of the hypolimnion.

2.7. Nitrogen and Phosphorus Cycles

All systems need two major chemical components for the growth, reproduction, and general metabolic processes of the organisms that inhabit those systems. These components, termed essential elements, have been phosphorus and nitrogen. Plants need phosphorus and nitrogen in their dissolved form as inorganic phosphate (termed orthophosphate and abbreviated *o* PO_4) and inorganic nitrate (NO_3). Although animals also need phosphorus and nitrogen, they need it in its organic form as organic phosphorus and organic nitrogen (generally obtained in the form of amino acids or protein by grazing on plants or predation on other consumers). Should animals attempt to ingest phosphate or nitrate in its inorganic form, it would be totally unusable and immediately excreted.

Phosphorus and nitrogen tend to cycle (Fig. 2.4) in systems, passing from plant to various animal communities, then into their inorganic forms, and then back into the plant components of the system. The discussion of these cycles is conventiently begun with those plants which will take in the inorganic materials.

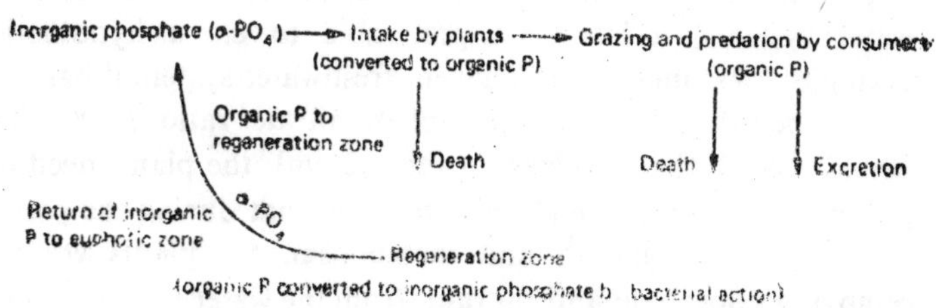

Fig. 5.4. The Phosphorus cycle.

During the process of photosynthesis the plants are able to convert o-PO_4 and NO_3 into their organic forms like nucleic acids (DNA, RNA), amino acids, and proteins, which contain various ratios of phosphorus and nitrogen in their molecular structures. These materials are used by the plants in their metabolic process and stored as new plant tissue during the growing season. Herbivores (primary consumers) get their phosphorus and nitrogen by feeding directly on the plants in the system, while the higher consumers obtain their phosphorus and nitrogen by predation. At each step in the food chain some of the material gets lost to the next link by death and/or excretion. In either case the phosphorus and nitrogen present in the excretory matter or the bodies of the plants and/or animals not consumed will sink through the water column and enter the benthic zone. Although bacterial decomposition presumably starts from the moment of excretion or at death, the majority of the decomposition occurs in the benthic sediments (regeneration zone). It has been in the regeneration zone that bacterial action converts the organic nitrogen and phosphorus into their inorganic form. It has to be noted (Fig. 2.4) that phosphorus readily gets converted into its inorganic form.

Plants inhabiting any system need nitrogen and phosphorus in certain ratios. The ratios needed by plants in both terrestrial and freshwater systems have been extremely variable and thus are dependent upon the specific populations of plants that inhabit a particular system. Since, however, plant communities do require not only phosphorus and nitrogen but need these materials in certain ratios, it is the availability of these nutrients in their proper ratio that becomes important from the viewpoint of productivity. As these ratios have been so variable in freshwater systems, it has been preferable to cite a hypothetical example. Assume that in a given freshwater system it has been determined that the proper phosphate/nitrate ratio is 1.8. This would reveal that (at least in this system) the plants need one phosphate for every eight nitrates, and that during the periods of active plant growth and metabolism, the plants would be removing phosphate and nitrate from the water in a 1 : 8 ratio and tying it up, as plant tissue, in this ratio. The consumers, too, would be getting the same phosphorus/nitrogen ratio. When these organisms died and fell to the regeneration zone, the decomposers would, ultimately, return this material to the water column in this ratio.

In a system in which the producers need a 1 : 8 ratio of phosphate to nitrate, should the ratios become disturbed (1 : 16, for example the plants would tend to remove 8 nitrates and all phosphate. The remaining 8 nitrates could not be used as there would not be sufficient phosphate in the system. In this case the factor limiting further plant growth would be phosphate. If the situation has been reversed and there were a 2 : 8, ratio, the limiting factor would be the lack of additional nitrates. In this case the plants would be able to remove all the nitrate but, since the nitrates were present in the lesser ratio, they would only be able to use one of the phosphates.

In many studies it is advisable to not only measure the inorganic phosphate and nitrate present in a given system, but also to determine the amount of nitrate, ammonia, and organic phosphate. By performing these additional analyse, not only can the productivity at the time of sampling by determined, but also the potential productivity in a given system can be inferred.

In the majority of freshwater systems it is found that phosphate has been generally the limiting factor. This has been due to the presence of blue green algae, which have been common in most freshwater systems. These forms have the ability to fix atmospheric nitrogen, thus increasing, the concentration of nitrate in the water column. Apparently the normal phosphate input into these systems from the weathering of phosphatic terrestrial materials is insufficient to counterbalance the amount of nitrogen input by the blue-green algae.

As it is known that both lentic and lotic systems, are having the same basic mechanisms of nutrient cycling and sunlight relationships as noted above. Beyond these commonly shared characteristics, however, the systems very greatly and should be considered as separate and discreate entities.

2.8 Complexation in Natural Water and Waste Water

Naturally occuring chelating agents like humic acids and amina acids are found in natural water and soil. The abundance of chloride in sea water gives rise to the formation of some chloro-complexes. There are synthetic chelating agents, like sodium tripolyphate, sodium ethylenediamine tetracetate (EDTA), sodium nitiriotriacetate (NTA) and sodium citrate, which are discharged in small amounts into aquatic system from different industrial wastes.

The ligands occuring in natural water and waste water are having a variety of organic groups which are able to co-ordinate to metal ions :

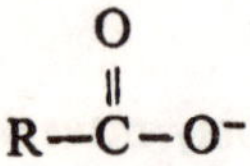

Carboxylate group

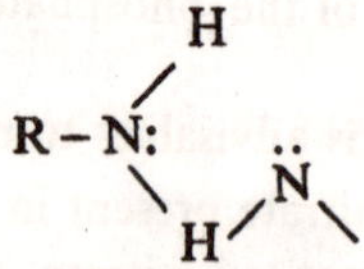

aliphatic and aromatic amino group

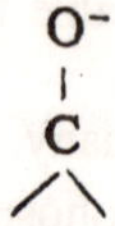

phenoxide group

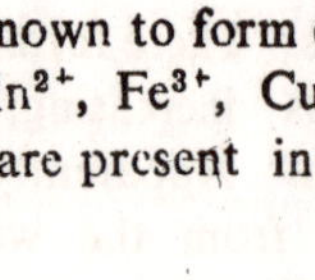

Phosphate group

These ligands are known to form complexes with most metal ions (Mg^{2+}, Ca^{2+}, Mn^{2+}, Fe^{3+}, Cu^{2+}, Zn^{2+}, CO^{2+}, Ni^{2+}, Sr^{2+}, Cd' and Ba^{2+}) which are present in natural water and biological system.

Humic Substances

These are the complexing agents which are important and found in nature. Although their existence has been known since 1800, yet their structural and chemical properties have been still challenging chemists.

Humic substances have been non-degradable materials which are formed during the decomposition of vegetation. They are found as deposits in soil, marsh sediments and any location where there have occurred large quantities of vegetation.

They are generally classified on the basis of solubility. If a humic substance could be extracted with a strong base and the resulting solution acidified the products have been (a) a non-extractable plant residue termed as humin, (b a material which gets precipitated from the acidified extract, termed as humic acid and (c) an organic material which is left in the acidified solution, called fulvic acid. The properties of water have been mainly affected by the humic substances, both soluble and insoluble because of their acid-base adsorptive and complexing properties. The soluble fulvic acid is having an effect on the properties of water, while the insoluble humin and humic acids are able to alter water quality by involving exchange of cations, organic materials, etc., with water.

The elementary composition of humic substances, C, 45–55%; O, 30 45%; H, 3–6%; N, 1–5% and S, 0–1% 'Humin', humic acid, and 'fulvic acid', are not applicable to single compounds but to a wide range of compounds. In general, humic substances constitute the high molecular-weight polyelectrolytic macromolecules whose molecular weights range from a few hundred (fulvic acids) to tens of thousands (humic acid and humin fractions).

They are having a carbon skeleton with a high degree of aromatic character and a large percentage of functional groups having oxygen. They may also be having protein-like materials and a carbohydrate fraction. These fractions can readily get hydrolyzed from the aromatic nucleus which is able to withstand chemical and biochemical attack.

When these occurs the decomposition of humic acid, the following typical decomposition products are obtained :

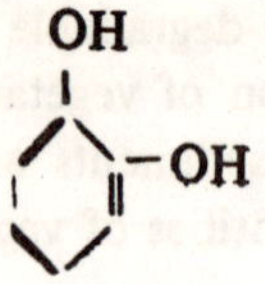

Catechol

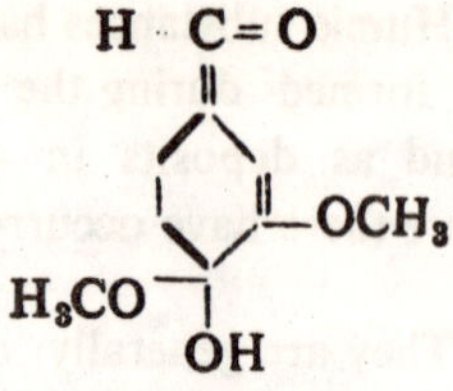

Syringaldehyde

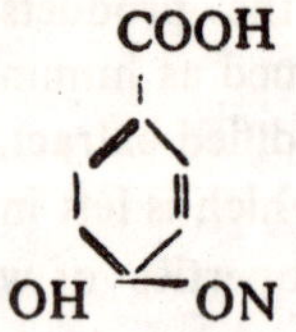

3, 5-Dihydroxy benzoic acid

The hypothetical structural formula of fulvic acid may be put as follows :

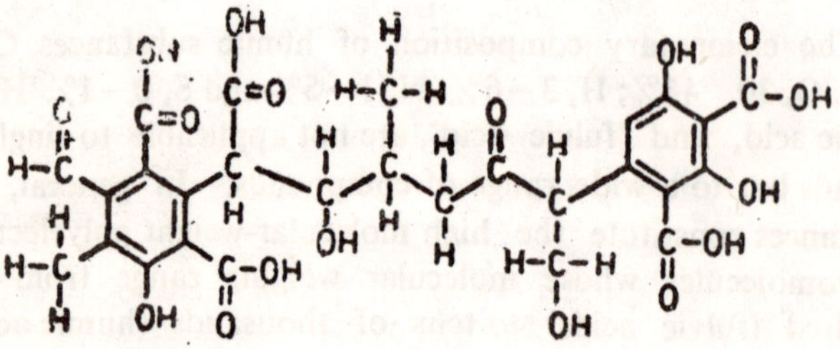

Fulvic acid

The fulvic acid is having a formula weight 666 and chemical formula, $C_{20}H_{15}$ $(COOH)_6$ $(OH)_5$ $(CO)_2$.

Humic substances are able to chelate metal ions through a carboxyl and a phenolic hydroxyl group.

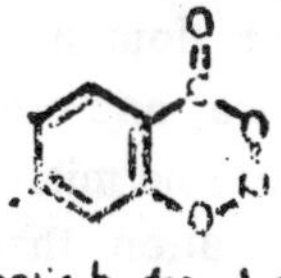

phenolic hydroxyl group

with two carboxyl groups.

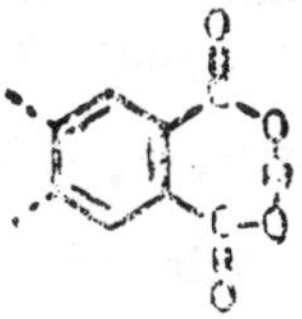

carboxyl group

or with a carboxyl group,

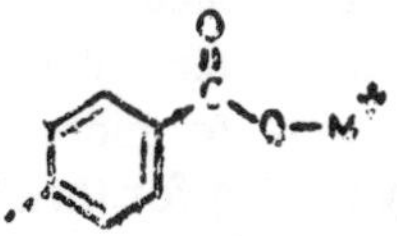

carboxyl group

Insoluble humic substances such as the humic and humic acids are able to exchange cations with water and are having the capacity to accumulate large amounts of metals.

Micro-Organic – The Catalyst of Aquatic Chemical Reactions

Micro-organisms such as the bacteria, fungi, and algae have been living catalysts which are able to being about a vast number of chemical processes in water and soil. A majority of the important chemical reactions taking place in water, particularly those involving organic matter and oxidation-reduction processes, take place through bacterial intermediaries. Algae constitute the primary produces of biological organic matter (biomass) in water. Micro-organisms have been considered to be responsible for the formation of many sediment and mineral deposits. Algae photosynthesis increases the pH of water thus causing the formation of calcium carbonate deposits. Some anaerobic bacteria produce hydrogen sulfide which reacts with metal ions to produce sulfide mineral deposits. With present technology, micro-organisms are able to play the dominant role in secondary waste treatment.

Pathogenic micro-organism must be eliminated from water purified for domestic use because the major epidemics of typhoid, cholera, and other water borne diseases have been resulted from pathogenic micro-organisms in water supplies. There is some speculation that viruses in water escape conventional water treatment and cause respiratory and digestive tract diseases, hepatitis, and perhaps even some forms of cancer.

Types of Micro-organisms in Waters

In considering aquatic chemistry, micro-organisms may be included in the three categories of bacteria, fungi, and algae. Fungi and bacteria (with the expception of photosynthetic bacteria) are termed as reducers. Reducers break down chemical compounds to more simple species, and thereby extract the energy needed for their growth and for their metabolic needs. As reducers can utilize only chemical energy, any chemical transformation mediated by them must involve a net loss of free energy. However, compared to higher organisms, the energy utilization of bacteria and fungi has been very efficient.

Algae have been classified as producers, as they use light energy and store it as chemical energy. In the absence of sunlight, however, algae use chemical energy for their metabolic needs. In a sense, therefore, bacteria and fungi may be looked upon as environmental catalysts, while algae are aquatic "solar fuel cells." Some of the chemical relationships among micro-organisms have been depicted in Fig. 2.5.

Algae may be regarded as generally microscopic organisms which subsist on inorganic nutrients and produce organic matter from carbon dioxide by photosynthesis. The general nutrient requirements of algae are carbon (from CO_2 or HCO_3^-), nitrogen (generally as NO_3^-), phosphorus (as some form of orthophosphate), sulfur (as SO_4^{2-}), and trace elements such as sodium, potassium, calcium, magnesium, iron, cobalt, and molybdenum.

In a highly simplified form, the production of organic matter by algae photosynthesis may be put by the equation

$$CO_2 + H_2O \xrightarrow{h\nu} (CH_2O) + O_2 \uparrow \quad ...(1)$$

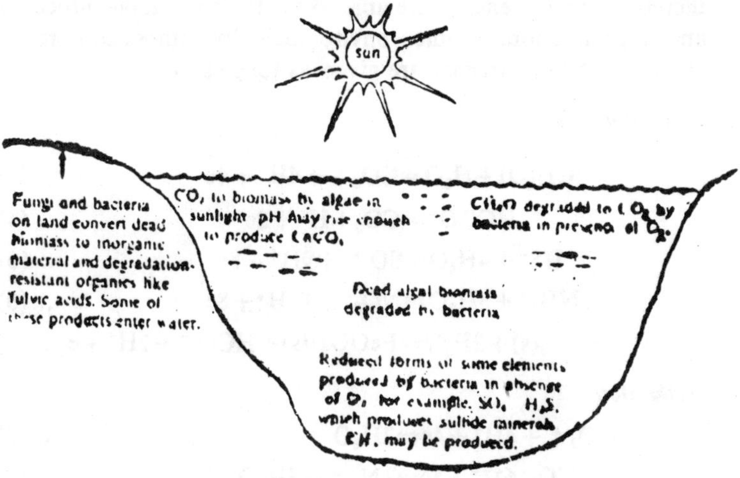

Fig. 2.5. Effects of micro-organisms on the chemistry of water.

.*/*here CH_2O denotes a unit of carbohydrate and *hν* stands for the energy of a quantum of light. Fogg (1) has represented the overall formula of the algae, *Chlorella*, as $C_{5.7}H_{9.8}O_{2.3}N$ (including phosphorus the formula would be $C_{5.7}H_{9.8}O_{2.3}NP_{0.06}$). using Fogg's formula for algae biomass, the overall reaction for photo-synthesis may be the following :

$$5.7CO_2 + 3.4H_2O + NH_3 \xrightarrow{h\nu} C_{5.7}H_{9.8}O_{2.3}N + 6.25O_2 \quad .. (2)$$

In the absence of light, algae are able to metabolize organic matter in the same manner as do nonphotosynthetic organisms. Thus, algae may be ab'e to satisfy their metabolic demands by using chemical energy from the degradation of stored starches or oils, or from the consumption of algae protoplasm itself In the absence of photosynthesis the metabolic process uses oxygen, so that during the hours of darkness an aquatic system having a heavy growth of algae may get depleted in oxygen.

Microbially Mediated Redox Reactions

Bacteria are able to mediate a variety of redox reactions and thereby drive the energy are able to for their metabolic processes and reproduction. Some environmentally important redox reactions of this category are given as follows ;

Oxidation

$$\{CH_2O\} + H_2O \rightleftharpoons CO_2 \uparrow + 4H^+ + 4e \quad ..(3)$$

$$HCOO^- \rightleftharpoons CO_2 + H^+ + 2e \quad ..(4)$$

$$HS^- + 4H_2O \rightleftharpoons SO_4^{2-} + 9H^+ + 8e \quad ..(5)$$

$$NH_4^+ + 3H_2O \rightleftharpoons NO_3^- + 10H^+ + 8e \quad ...(6)$$

$$FeCO_3(s) + 2H_2O \rightleftharpoons FeOOH(s) + HCO_3^- + 2H^+ + e \quad ..(7)$$

Reduction

$$O_2 \uparrow + 4H^+ + 4e \rightleftharpoons 2H_2O \quad ..(8)$$

$$NO_3^- 6H^+ + 5e \rightleftharpoons \tfrac{1}{2}N_2 \uparrow + 3H_2O \quad ...(9)$$

$$CO_2 \uparrow + 8H^+ + 8e \rightleftharpoons CH_4 \uparrow + 2H_2O \quad ...(10)$$

Bacteria have been taking fact in many biogeochemical processes in water and soil and in many important elemental cycles in nature, such as those of nitrogen, carbon and sulphur. They help in the formation of many mineral deposits including some of those of iron and manganese. Bacteria have been also instrumental in the formation of iron and manganse deposits in natural water systems and pipes used for transporting water.

Nitrogen Fixation

The biochemical pathway of nitrogen fixation by which atmospheric N_2 is chemically bound in biological or simple inorganic compounds is not simple, and a number of conflicting theories have been advanced to explain the mechanisms of the process. Tracer studies of this vital process by which atmospheric nitrogen gets incorporated into living matter revealed that the first detectable fixed nitrogen species has been NH_3. The ammonia has been then converted rapidly to glutamic and aspartic acids by involving the following reactions ;

$$HO_2C-\overset{\overset{\displaystyle O}{\|}}{C}-CH_2-CH_2-CO_2H+NH_4^{+}+H^{+}+2e=$$
α-Ketoglutaric acid

$$^{(-)}O_2C-\overset{\overset{\displaystyle NH_3^{(+)}}{|}}{CH}-CH_2-CH_2-CO_2H+H_2O \qquad \ldots(1)$$
glutanic acid

$$HO_2C-CH=CH-CO_2H+NH_3={}^{(-)}O_2C-\overset{\overset{\displaystyle NH_3^{(+)}}{|}}{CH}-CH_2CO_2H$$
Fumaric acid — aspartic acid ...(2)

Few kinds of aquatic micro-organisms are having the ability to fix atmospheric nitrogen. Among the aquatic bacteria which are having this capability have been photosynthetic bacteria Azotobacter, and several varieties of Clostridium. Among the algae, blue-green algae are able to fix atmospheric nitrogen. In most natural fresh water systems, however, the fraction of nitrogen fixed by organisms in the water relative to that originating from the decay of organic material, fertilizer runoff, and other external sources, has been quite low.

The best known and most important form of nitrogen-fixing bacteria has been Rhizobium, which has a symbiotic (mutually advantageous) relationship with leguminous plants such as clover or alfalfa. The Rhizobium bacteria occur in root nodules, special structures attached to the roots of legumes. These bacteria are able to fix atmospheric nitrogen in a chemical form which the legumes can utilize. When the die and decay, some of the nitrogen gets converted by micro-organisms in the decay process to nitrate ion, which may be utilized by other plants. Before the advent of commercial chemical fertilizers, the growth of leguminous crops was widely used as a means of building soil fertility. Some of the nitrate produced by the decay of legumes enters natural waters. In an indirect way, therefore, Rhizobium bacteria have been responsible for the introduction of some nitrogen into waters.

Nitrification

Nitrification, the conversion of N(– III) to N(V), has been a very common and extremely important process in water and in soil. As it is known aquatic nitrogen in thermodynamic equilibrium with air has been in the form of NO_3 ion, N(V), whereas in most biological compounds nitrogen occur as N(– III), such as $—NH_2$ in amino acids. The equilibrium constant for the overall nitrification reaction, written as one electron-mole,

$$\frac{1}{4}O_2 + \frac{1}{8}NH_4^+ \rightarrow \frac{1}{8}NO_3^- + \frac{1}{4}H + \frac{1}{8}H_2O \quad .. (1)$$

is $10^{7.69}$ so the reaction is highly favored from a thermodynamic viewpoint.

One of the most important ramifications of this conversion has been that nitrogen gets absorbed by plants primarily as nitrate. When fertilizers are applied in the form of ammonium salts or anhydrous ammonia, a microbial transformation to nitrate enables maximum assimilation of nitrogen by the plants.

If extensive aeration has been allowed to take place in the activated sludge sewage treatment process the ammonia nitrogen in the sewage is largely oxidized to nitrate ion. As the sewage sludge settles out in the settler, the bacteria in the sludge use this nitrate as an oxygen source producing N_2 gas. The bubbles of nitrogen gas bring about the sludge to rise, so that it does not settle properly. This can hinder the proper treatment of sewage through carryover of sludge into effluent water.

In nature, nitrification gets catalyzed by two groups of bacteria, Nitrosomonas and Nitrobacter. Nitrosomonas bacteria cause the transition of ammonia to nitrite,

$$NH_3 + \frac{3}{2}O_2 \rightarrow H^+ + NO_2^- + H_2 \quad (2)$$

whereas Nitrobacter mediate the oxidation of nitrite to nitrate :

$$NO_2^- + \frac{1}{2}O_2 \rightarrow NO_3^- \quad .. (3)$$

Both of these highly specialized types of bacteria are obligate aerobes, that is, they function only in the presence of molecular O_2. These bacteria are also chemolithotrophic, implying that they can use oxidizable inorganic materials as electron donors in oxidation reactions to yield needed energy for metabolic processes. A majority of chemolithotrophic bacteria use CO_2 as a carbon source in the synthesis of their biomass and have been therefore, autotrophic. There have been several classes of chemolithotrophic bacteria besides the nitrifying bacteria. These other classes include hydrogen bacteria, colorless sulfur bacteria, and iron bacteria.

For the aerobic conversion of ammoniacal nitrogen to nitrite ion at pH 7.00,

$$\frac{1}{6}\,NH_4^+ + \frac{1}{4}O_2(g) \rightarrow \frac{1}{6}NO_2^- + \frac{1}{3}\,H^+ + \frac{1}{6}H_2O \quad ..\ (4)$$

the free energy change is -10.8 kcal. The free energy change for the aerobic oxidation of one electron mole of nitrite ion to nitrate ion

$$\frac{1}{2}\,NO_2^- + \frac{1}{4}O_2(g) \rightarrow \frac{1}{2}NO_3^- \quad ...(5)$$

has been -9.0 kcal. Both steps of the nitrification process involve an appreciable yield of free energy. It has been interesting to note that the free energy yield per electron mole has been approximately the same for the conversion of NH_4^+ to NO_2^- as it is for the conversion of NO_2^- to NO_3^-, about 10 kcal/electron-mole.

Nitrate Reduction

The general term "nitrate reduction" means the microbial processes by which nitrogen in chemical compounds gets reduced to lower oxidation states. In the absence of free oxygen, nitrate may be used by some bacteria as an alternate electron receptor. The most complete possible reduction of nitrogen in nitrate ion involves the acceptance of 8 electrons by the nitrogen atom with the consequent conversion of nitrate to ammonia (+V to −III oxidation state). Nitrogen is an essential component of protein, and any organism which utilizes nitrogen from nitrate for the synthesis of protein must first reduce the nitrogen to the −III oxidation state (ammoniacal form). However, incorporation of nitrogen into protein generally has been a relatively minor use of the nitrate undergoing microbially-mediated reactions, and has been more properly termed nitrate assimilation.

Generally when nitrate ion functions as an electron receptor, the product has been NO_2^-. When nitrate ion acts as an electron receptor,

$$\frac{1}{2}NO_3^- + \frac{1}{4}CH_2O = \frac{1}{2}NO_2^- + \frac{1}{4}H_2O + \frac{1}{4}CO_2$$

the free energy yield per electron mole has been only about 2/3 of the yield when oxygen is the electron receptor. However, nitrate ion is a good electron receptor in the absence of molecular oxygen. One of the factors limiting the use of nitrate ion in this function has been the relatively low concentration of NO_3^- in most waters. Furthermore, nitrite, NO_2^-, has been relatively toxic and tends to inhibit the growth of many bacteria after building up to a certain level. Sodium nitrate has been sometimes used as a "first-aid" treatment in sewage lagoons which have become oxygen-deficient. It provides an emergency source of oxygen to re-establish normal bacterial growth.

Deintrification

An important special case of nitrate reduction has been denitrification, in which the reduced nitrogen product has been gaseous N_2. At pH 7 00 the free energy change per electron-mole of reaction

$$\frac{1}{5}NO_3^- + \frac{1}{4}CH_2O + \frac{1}{5}H^+ = \frac{1}{10}N_2(g) + \frac{1}{4}CO_2(g) + \frac{7}{20}H_2O \quad \text{.. (1)}$$

is −2.84 kcal. The free energy yield per mole of nitrate reduced to N_2 (5 electron-moles) has been higher than that for the reduction of the same quantity of nitrate to nitrite. More important, however, the reduction of a nitrate ion to N_2 gas consumes 5 electrons, compared to only 1 electron for the reduction of NO_3^- to NO_2^-.

Deintrification has been an important process in nature. It is the mechanism by which fixed nitrogen gets returned to the atmosphere. Deintrification is also useful in advanced water treatment for the removal of nutrient nitrogn. The water is treated with a minimum amount of methanol under anaerobic conditions, and N_2 gas is evolved according to the following reaction :

$$5CH_3OH + 6NO_3^- + 6H^+ = 5CO_2 + 3N_2\uparrow + 13H_2O \quad \text{.. (2)}$$

As nitrogen gas is a nontoxic volatile substance which does not inhibit microbial growth, and as nitrate ion is a very efficient electron acceptor, denitrification allows the extensive growth of bacteria under anaerobic conditions.

The Nitrogen Cycle

The relationships described in the preceding sections on nitrogen transformations of bacteria may be summarized in the nitrogen cycle. The nitrogen cycle describes the dynamic processes through which nitrogen gets interchanged among the atmosphere, organic matter, and inorganic nitrogen compounds.

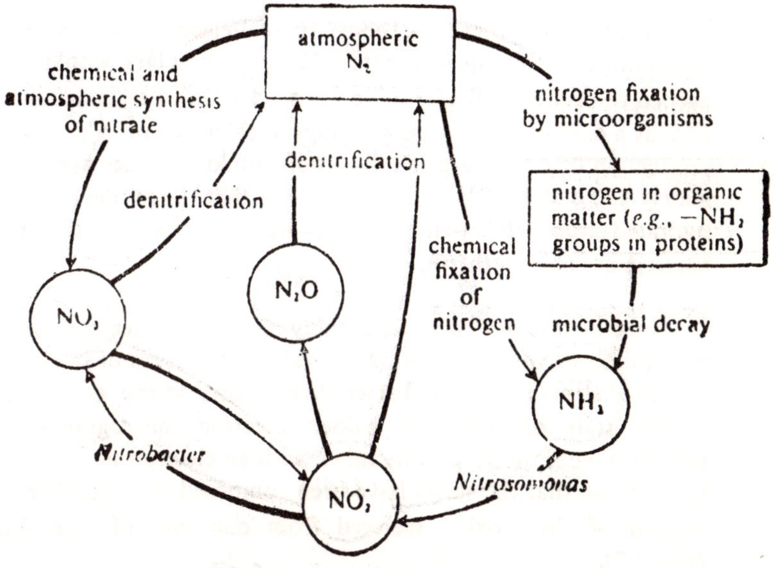

Fig. 2.6. The nitrogen cycle.

It is one of nature's vital dynamic processes. The nitrogen cycle is shown in Fig. 2.6.

Iron and Manganese Bacteria

Some bacteria, including Ferrobacillus, Gallionella, and some forms of Sphaerotilus, use iron compounds in getting energy for their matabolic needs. The bacteria are able to catalyze the oxidation of iron (II) to iron (III) by molecular oxygen :

$$4Fe(II)+4H^{+}+O_2 \rightarrow 4Fe(III)+2H_2O \qquad ...(1)$$

The carbon source for some of these bacteria is CO_2. As they do not need organic matter for carbon, and because they derive energy from the oxidation of inorganic matter, these bacteria may be able to thrive in environments where organic matter is absent.

The micro-o ganism mediated oxidation of iron (II) has been not a particularly efficient means of energy for metabolic processes. For the reaction

$$FeCO_3+\frac{1}{4}O_2+\frac{3}{2}H_2O=Fe(OH)_3+CO_2 \qquad ...(2)$$

the change in free energy is almost 10 kcal/electron-mole. It has been estimated that approximately 220 g of iron (II) would be needed to produce 1.0 g of cell carbon. The calculation assumes CO_2 as a carbon source and a biological efficiency of 5%. The production of only 1.0 g of cell carbon would produce approximately 430 g of $Fe(OH)_3$. It is evident that large deposits of hydrated iron (III) oxide form in areas where iron oxidizing bacteria are able to thrive.

Some of the iron bacteria, notably Gallionella, secrete large quantities of hydrated iron (III) oxide in the form of intricate branched structures. The bacterial cell grows at the end of a twisted stalk of the iron oxide. Electron micrographs of individual cells of Gallionella taken with an electron microscope have shown that the stalks have been composed of a number of strands of iron oxide secreted from one side of the cell (Fig. 2.7).

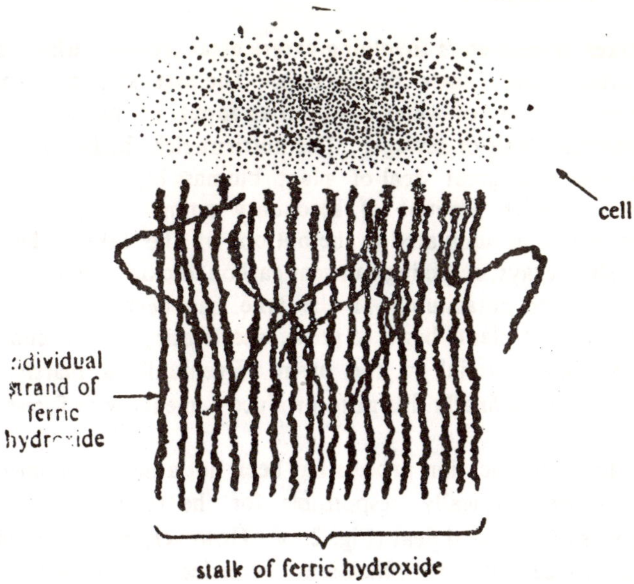

Fig. 5.7. Sketch of a cell of Gallionella showing ferric hydroxide secretion.

At pH values near neutrality, bacteria deriving energy by mediating the air oxidation of iron(II) must compete with direct chemical oxidation of iron(II) by O_2. The latter process has been relatively rapid at pH 7. As a consequence, these bacteria tend to grow in a narrow layer in the region between the oxygen source and the source of iron(II). Hence, iron bacteria are sometimes termed as "gradient organism," and they grow at intermediate pE values.

Bacteria have been strongly involved in the oceanic manganese cycle. Manganese nodules, a potentially important source of manganese, copper, nickel, and cobalt give rise to different species of bacteria which mediate both the oxidation and reduction of manganese. These reactions are enzymatic in nature and get influenced by seawater cations, especially Ca^{2+} and Mg^{2+}.

Eutrophication

The term "eutrophication" has been derived from Greek word which means "well nourished". It describes a condition of lakes or reservoirs involving excess algal growth, which may eventually cause their destruction. The first step in eutrophication of a body of water has been an input of nutrients from watershed runoff or sewage. The nutrient-rich body of water gives rise to a great deal of plant biomass by photosynthesis, and a smaller amount of secondary animal biomass. Deal biomass gets accumulated in the bottom of the lake, where it partially decays, recycling nutrient carbon dioxide, phosphorus, nitrogen, and potassium. If the lake has been not too deep, bottomrootéd plants begin to grow, accelerating the accumulation of solid material in the basin. Eventually a marsh gets formed, which finally fills in to produce a meadow or forest.

The eutrophication process has been not a new phenomenon. It has been basically responsible for the formation of huge deposits of coal and peat (Fig. 2.8). However, man's activities are able accelerate eutrophication. In order to understand why this is so, refer to Table 2.2. This table shows the chemical

Fig. 2.8. The eutrophication process, occurring millions of years ago was responsible for the formation of huge deposits of coal and other fossil fuels.

elements needed for plant growth. Most of these occur at a level which is more than sufficient to support plant life in the average lake or reservoir. Hydrogen and oxygen are provided by the water itself. Carbon is provided by CO_2 from the atmosphere or from decaying vegetation. Sulfate, magnesium, and calcium normally occur in abundance from mineral strata in contact with the water. The micronutrients are needed at only very low levels (for example, approximately 40 ppb for copper).

TABLE 2.2

Essential Plant Nutrients : Sources and Function

Nutrient	*Source*	*Purpose*
Macronutrients		
carbon (CO_2)	atmosphere, decay	biomass constituent
hydrogen	water	biomass constituent
oxygen	water	biomass constituent
nitrogen (NO_3^-)	decay, atmosphere (nitrogen fixing organisms), pollutants	protein constituent
phosphorus (phosphate)	decay, minerals, pollutants	DNA, RNA constituent
potassium	minerals, pollutants	metabolic function
sulfur (sulfate)	minerals	proteins, enzymes
magnesium	minerals	metabolic function
calcium	minerals	metabolic function
Micronutrients		
B, Cl, Co, Cu, Fe, Mo, Mn, Na, Si, V, Zn	minerals, pollutants	metabolic function and or constituent of enzymes

Therefore, the nutrients most likely to be limiting have been the "fertilizer" elements : nitrogen, phosphorus, and potassium. These all occur in sewage and are of course found in runoff from heavily fertilized fields. They are also constituents of various kinds of industrial wastes. Each of these elements can come from natural sources phosphorus and potassium from mineral formations, and nitrogen fixed by bacteria, blue-green algae, or discharge of lightning through the atmosphere.

Generally the plant nutrient most likely to be limiting has been phosphorus, and it has been generally named as the culprit in excessive eutrophication. Household detergents constitute a common source of phosphate in wastewater, and eutrophication control has concentrated upon eliminating phosphates from detergents, removing phosphate at the sewage treatment plant, or preventing phosphate-laden sewage effluents (treated or untreated) from entering bodies of water.

In some cases nitrogen or even carbon constitute limiting nutrients. Particularly in sea water nitrogen may be a limiting nutrient.

The whole eutrophication picture has been a complex one and continued research is required to solve the problem. It is indeed ironic that in a food-poor world, nutrient-rich wastes from over-fertilized fields or sewage are bringing about excessive plant growth in many lakes or reservoirs. This illustrates again that in general there has been no such thing as a pollutant, there have been only resources (in this case plant nutrients) gone to waste.

Water Circulation

The three great water reservoirs of the earth - the seas, the continents and the atmosphere - are not in any sense self contained units but are in constant communication with one another. The relationships take the form of a circulation. Under the influence of solar radiation, water is constantly vaporized and emitted into the atmosphere in the form of water vapour. This process can occur in a great variety of ways. On the one hand, water evaporates directly from the surface of seas lakes and rivers, from glaciers and snowfields, or straight out of the ground. On the other hand, in the process of breathing, all living organisms, animals and plants alike, emit both carbon dioxide and water vapour into the atmosphere. A third process is the emission of water vapor by the combustion of organic materials such as wood, coal and oil, which are to be found all over the earth. Easily the highest proportion of vaporized water emanates directly from the oceans, however.

The water suspended in the atmosphere in the form of water vapour is of the greatest importance to the climate conditions on earth. It conditions the composition of the air masses, exerts an enfluence on the energy conditions of the atmosphere, and substantially governs and sustains the water circulation.

The water content in the atmosphere is measured in terms of the relative humidity of the air. Whereas some portions of the vaporized water masses remain permanently suspended in the atmosphere in the form of moisture, others are condensed back by the cooling down of ascending airstreams and emerge in the form of clouds, fog, rain or dew. If the water cools still further, snow or hailstones may build up in the clouds. When the clouds get too heavy, precipitation begins and the moisture returns to the earth in the form of rain, snow or hail. If the precipitation goes straight into the sea or into lakes having no outlet, the water circulation will be complete. If it falls on land, some of it will accumulate by surface drainage in brooks, streams and rivers and so return to the ocean. Other forms of precipitation seep into the earth and replenish the groundwater, or if retained at an upper ground level as trap water will evaporate from there. The trap water is also available to sustain plant life, which absorbs it through its roots and returns it to the atmosphere as water vapour by respiration and transpiration.

This circulation, like all others on the earth, cannot occur in the absence of some motive force. The entire cycle is maintained by the regular inflow of solar energy. Of the total amount of energy absorbed by the earth, abcut one-third is utilized in maintaining the water circulation. The temperature, humidity and movement (wind) in the atmosphere together determine the level of evaporation.

Although the total volume of atmospheric moisture is very small in comparison with the total hydrosphere, owing to the circulation of the atmosphere an enormous amount of water movement takes place in the course of a year. To produce the annual precipitation on the earth of about 470,000 cubic kilometres, the water in the atmosphere (about 12,300 cubic kilometers) must be replaced 35 to 40 times a year.

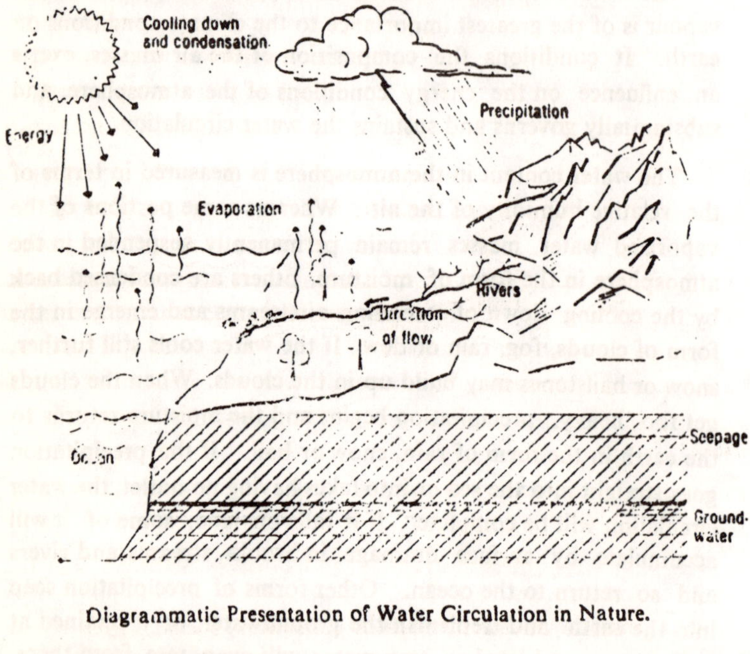

Diagrammatic Presentation of Water Circulation in Nature.

Composition of Seawater

Seawater is a complex solution of varions salts, trace elements and gases. There is a measure of equilibrium between the input of soluble matter by the rivers, elutriation (or removal of suspended particles) from the atmosphere by rain, and the loss by precipitation and absorption of nutrients. The bulk of the soluble matter is washed into the sea through the weathering and erosion of rocks from the solid crust of the earth. Much of this matter is soon precipitated to form new rock deposits on the seabed, which become stratified over geological periods and can form new mountains again as a result of pressure movements in the earth's crust.

In shorter periods, variations can be caused in the composition of sea-water by biological processes (occasioned by the temperature, for instance) and today in particular by the deposit of waste and toxic substances in the sea. In general, the quantity of inorganic material dissolved in seawater amount to 35 grams per kilogram of water. This represents a solution of 3.5%.

To come down to details, seawater contains the following elements (average concentration):

(milligrams per liter)

Substance	*Concentration*
1	2
sodium	10,600
chlorine	18,990
magnesium	1,350
sulfur	885
calcium	400
potassium	390
carbon (as carbonates and carbon dioxide)	23 - 28
strontium	8.1
bromine	65
boron	4.6
silicon	0.02 - 4
fluorine	1.5
nitrogen	0.02—0.8
rubidium	0.12

(*Contd.*)

1	2
aluminium	0.01
lithium	0.18
barium	0.05
iodine	0.05
phosphorus	0.0001 – 0.1
arsenic	0.003
iron	0.002 – 0.02
manganese	0.001 - 0.01
copper	0.001 - 0.01
zinc	0.005 - 0.01
lead	0.00003
selenium	0.00009
cesium	0.0004
uraninm	0.003
silver	0.0003
thorium	0.00005

Most of these elements are present in seawater in the form of salts, and a smaller portion in the form of dissolved gases. The most important of these are oxygen and carbon dioxide. The oxygen content fluctuates between 0 and 8.5 milliliters per liter. The high concentrations occur near the surface of the water, where the oxygen is in equilibrium with the oxygen in the atmosphere. A second source of the oxygen dissolved in the sea comes from the photosynthesis of phytoplankton, the drifting oceanic plant life. At great depths it may happen that (owing to the intense activity of bacteria and animals that absorb oxygen) the oxygen content of the seawater falls to zero. In many places, however, owing to the downward trend of the convection currents, a large quantity of dissolved oxygen from the surface water is carried to considerable depths in the sea.

An important part is played by carbon dioxide gas, which is present in considerable quantities in seawater. Since the basic ions of sodium, potassium and calcium are also present in large quantities in seawater, this facilitates the solution of a relatively high percentage of carbon dioxide. Carbon dioxide is one of the basic substances for photosynthesis (see Part Six). The plants living in the sea (especially plankton) take the carbon dioxide they need for photosynthesis directly out of the surrounding water. The second important effect of carbon dioxide in seawater is its function as a buffer (dissolved matter which keeps the pH value of the solution practically constant over wide areas). Normally the pH value of seawater lies between 7.5 and 8.4. As carbon dioxide and its associated marine chemicals are in equilibrium, the pH value of seawater is approximately constant as regards the input of both acids and bases.

The other gases from the atmosphere are dissolved in seawater in much lower concentrations than oxygen and carbon dioxide. The concentrations are as follows :

(milligrams per liter)

Substance	*Concentration*
nitrogen	between 8.4 and 15
helium	0.000005
neon	0.0001
argon	0.6
krypton	0.0003
xenon	0.00005

A particular phenomenon is the enrichment of seawater by trace elements that occur in considerable concentrations in living organisms. An example is provided by the ascidian (a tunicate), a small, mostly clinging animal that increases the element vanadium in seawater up to 50,000 times its normal concentration. It is known today that many other trace elements such as iodine, arsenic, nickel, zinc, titanium, chromium and strontium are also very much increased in the water by decaying organisms. Certain fish are known to contribute chromium, nickel, silver, tin and zinc. Similarly, some of the substances produced by humans get into the sea as waste substances. As such a buildup of substances in the food web takes some time, the damage caused by them is often not noted until too late. The increase of chlorinated hydrocarbons, for instance (especially pesticides such as DDT), reaches its maximum concentration in fish only after a period of 11 years from its production.

Also of great importance in its effects on life in the sea is the content of ammonium compounds, nitrates and phosphates. As these are absorbed by plants to build up their bodies, their content in seawater varies considerably according to the presence of plant life. The following concentrations in seawater have been established.

(milligrams per liter)

Substance	*Concentration*
ammonia	0.4 – 50
nitrate	1 – 600
nitrous oxide (particularly NO_2)	0 – 15
organically combined nitrogen	30 - 100
phosphates	1 – 100
organically combined phosphorus	1 – 30

Groundwater

A large part of the preciaitated water (rain, snow, dew, etc.) sinks into the ground and supplements the underground water reserves. To be specific, two forms of subterranean water supplies can be distinguished : water that percolates through from the surface, and groundwater proper. Where the rainfall is light, the precipitation penetrates by seepage only into the upper strata and remains suspended in the highest levels. This is particularly evident when the ground consists of peat or other very clayey material. The highest stratum is then described as a water retentive zone. If the precipitation is greater or the subsoil is very sandy or gritty, the seepage water penetrates the upper strata and joins the groundwater. In the lower lying underground strata the seepage water spreads mostly over an impenetrable rock layer and so provides the groundwater. The water-carrying ground layers are described as groundwater layers.

Particularly rich in groundwater are the underground strata of sand and grit, which mostly emanate from the Pleistocene epoch, as many small fissures open up between the individual soil particles; these fissures can be filled with water.

Connecting lines between groundwater levels of the same size are commonly described as groundwater equivalents. From these it is seen that the groundwater is in constant interchange with the rivers. As a general rule the groundwater level is higher than the neighboring riverbed. The groundwater layer follows the line of the riverbed, though considerably more slowly, as the river flows downstream, and forms a reservoir for the river; when the river water level is low it is replenished from the groundwater, and when the water level is high it releases surplus water to the groundwater reserves.

In dry regions the groundwater lies considerably deeper than in wet regions. The seepage-water zone contains only ground moisture and as occasion offers also groundwater, which overcomes the force of gravity to rise by capillary attraction. As a result of the evaporation of the capillary waters on the surface, the minerals dissolved in the water are precipitated as salt and gypsum at on in the highest level.

In the bedrock a distinction can be made between quarry water and groundwater. Quarry water seeps through small hollow "reamers", or holes, in the bedrock, and serves gradually to enlarge them by physical and chemical processes. The chemical processes are particularly marked with a salt and gypsum subsoil or a limestone rock, owing to the high arbon dioxide content of the seepage water. Ground-stratum water is found primarily in sedimentary rock such as sandstone and loess. If impermeable strata are above or below, big subterranean water systems can be formed that can be tapped by artesian wells.

As a rule, groundwater is very rich in soluble matter. The concentration depends on the duration of the subterranean flow of water, the temperature, the solubility and composition of the rocks. Owing to the usually low oxygen content, oxidation processes occur in these layers only to a limited extent, so that the solubility of many componuds is retained. If such waters come again to the surface to be combined with oxygen, the dissolved compounds will be precipitated as oxides.

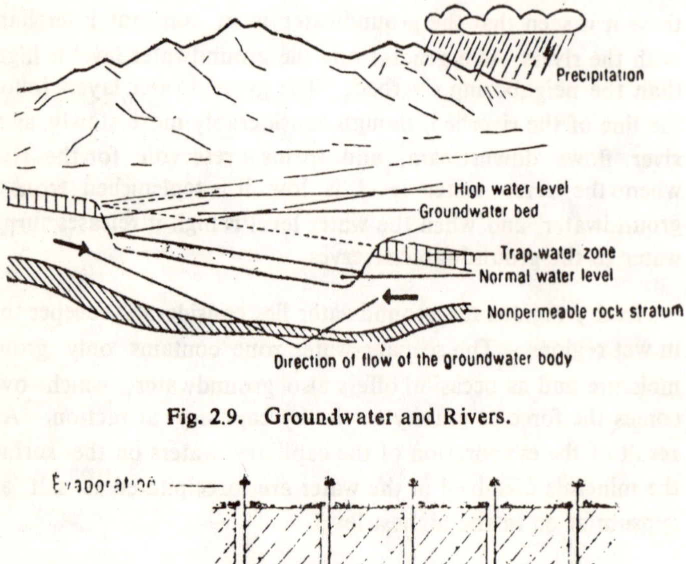

Fig. 2.9. Groundwater and Rivers.

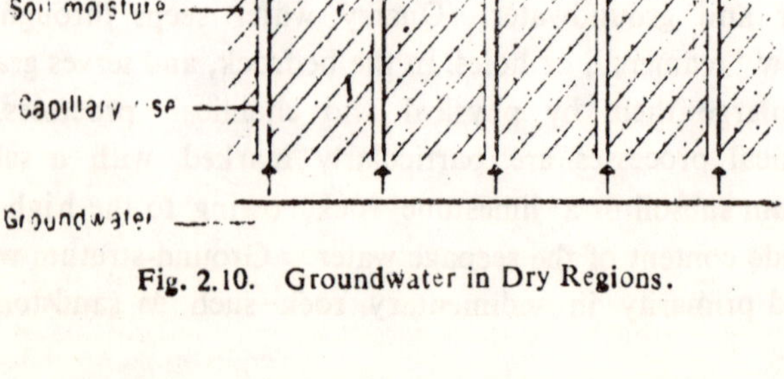

Fig. 2.10. Groundwater in Dry Regions.

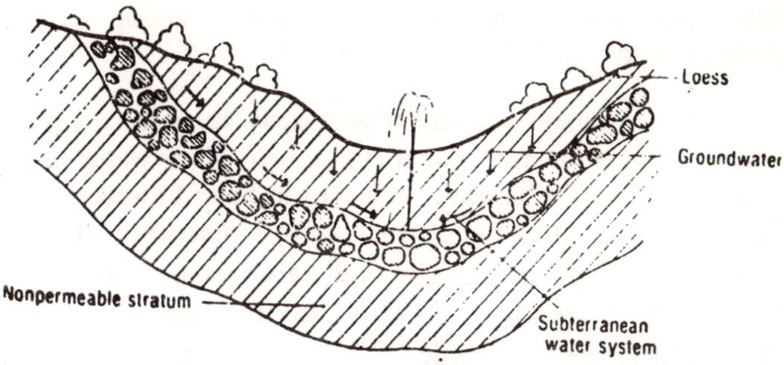

Fig. 5.11. Formation of Artesian Wells.

Heating of Rivers

Most power stations use river water for cooling purposes. This can give rise to far-reaching adverse ecological changes. The main problems are a reduction in the self purification capacity of the river and the danger to river fauna caused by the heating of the river water, the extermination of water organisms in the condenser in the power station, and the climatic effects of heating of the waters.

By the heating of river waters, the dissolving power for the oxygen that is vital to the needs of water organisms is lowered, and at the same time the metabolic processes of the water organisms are accelerated. A higher consumption of oxygen is thus accompanied by a reduction in the supply. This can result in an impairment of self-purification capacity, in the disappearance of oxygen (with the consequent extinction of water organisms and the possible extermination of fish life) and, in extreme cases, in the complete disruption of the water biology. The heating of river waters leads to a sharp increase in bacteria. One result of this is that in addition to the physical reduction in the oxygen content of the water already caused by the rise in temperature, the remaining oxygen is consumed more quickly. The water acquires a musty smell and taste, which can seriously complicate the purification of drinking water from the river concerned. The bacterial cultures that develop can settle in quiet spots in the river, such as backwaters beside weirs, and putrefy there.

As this problem was recognized fairly early on, authorities imposed limits on the extent to which river water might be heated. These limits have been set too high, however, as they only take account of the physical factors present in clean rivers. The limits laid down stipulate that water temperatures may not be raised by more than 5°C (41°F), so that the upper limit of 28°C (82°F) is never exceeded. This threshold value takes no account of the fact that this range of temperature can give rise, in polluted rivers, to extensive disturbances in ecological processes to the point of complete disruption. Detailed investigations have been made into the matter on the river Main in West Germany, from which it emerges that once the level of 23°-24°C (73°-75°F) is overstepped, organisms die from lack of oxygen caused by pollution and heating of the water.

A further problem consists in the sharp increase in temperature of the water drawn into the condenser at a power station, possibly to 40°C (104°F). This is precisely the optimum temperature for pathogenic bacteria such as typhus and enterococci (intestinal bacteria). In clean cold water these disease germs do not multiply, but in polluted warm water well supplied with nutrients they do so freely. Another serious factor is that when the water heats to 40°C in the condenser, the animals that prey on bacteria rotifers (wheel animalcules) and ciliates (slipper animalcules, etc.)—die off. An increase in pathogenic bacteria is thus matched by the extermination of the creatures that prey on them. In a 1000-megawatt power station, about 40,000 liters of water per second are absorbed and heated, which seriously impairs the self-purification capacity of polluted rivers.

A third problem arises from the influence on the climate of heating the river. Scientists at the Nuclear Research Center in Karlsruhe, Germany, have investigated this thoroughly as regards the Rhine. They reached the conclusion that with a rise in temperature of no more than 3°C (37.4°F) the quantity of

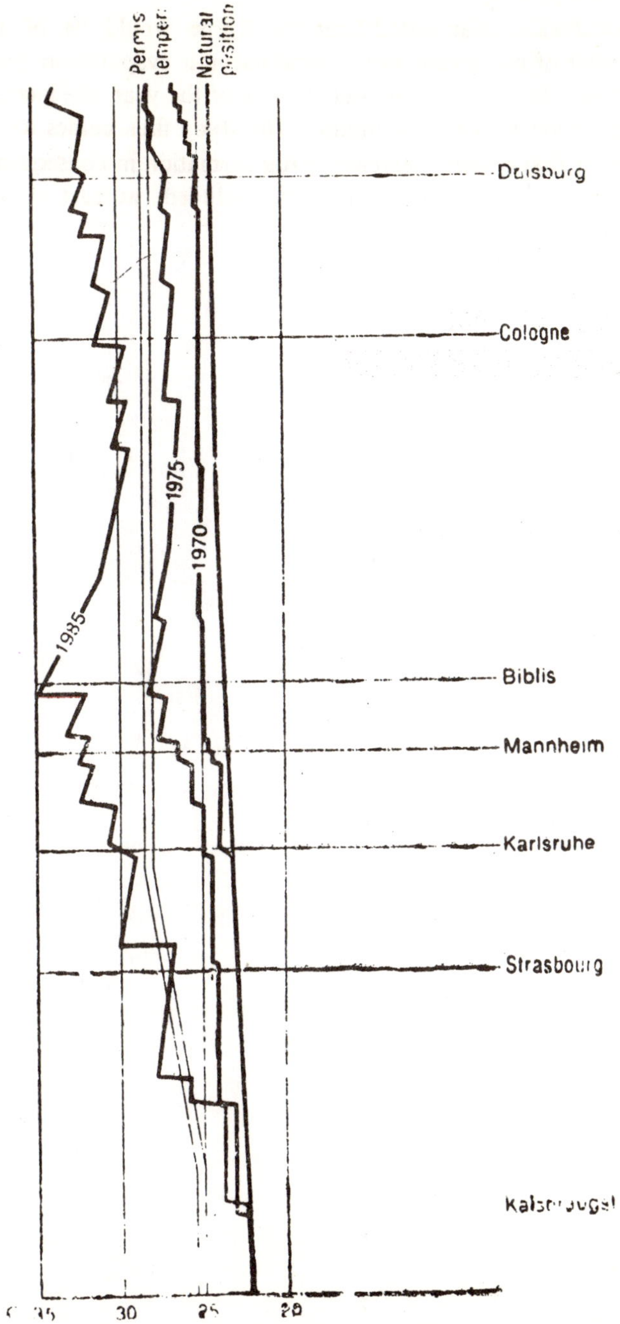

Fig. 2.12 Temperature Forecasts for the Rhine under the Summer-heat Load Plan

additional water evaporated from the Rhine would be of the same order of magnitude as the total natural evaporation from the Rhine. In the foggy winter months of the year, the increase in evaporation is at its maximum. The study thus comes to the conclusion that a sharp increase in fog formation in consequence of heating of the river cannot be excluded, at least in the proximity of the river.

3

Classification of Water Pollutants

Introduction

The signs of water pollution have been obvious to even the most casual observer. To aid in a systematic discussion of water pollutants, they will be classified into 9 categories given below. Each category is discussed in this chapter and then some specific pollutant have been studied in more detail in the following chapters.

1. Oxygen-demanding wastes
2. Disease causing agents
3. Plant nutrients
4. Synthetic organic compounds
5. Oil
6. Inorganic chemicals and mineral substances
7. Sediments
8. Radioactive materials
9. Heat (Thermal Pollution).

1. Oxygen Demanding Wastes

Dissolved oxygen DO has been a fundamental requiremental of life for the plant and animal population in any given body of water. Their survival is dependent upon the ability of the water to maintain certain minimal concentrations of this vital substance.

A body of water has been classified as polluted when the DO concentration drops below the level necessary for sustaining a normal biota for that water. The primary cause of water deoxygenation has been the presence of substances collectively called oxygen-demanding wastes. These are substances easily broken down or decayed by bacterial activity in the presence of oxygen. The available dissolved oxygen has been consumed by bacterial activity, and thus, the presence of such materials quickly gives rise to depletion of dissolved oxygen.

Oxygen-demanding wastes are primarily organic materials that are oxidized by bacteria to carbon dioxide and water. These substances have been deleterion because their decomposition gives rise to oxygen depletion in both freshwater and marine systems. They also produce unaesthetic odors, endanger water supplies, and decrease the recreational value of waterways.

Sewage and other oxygen demanding wastes have been classified as water pollutants because their degradation leads to oxygen depletion, which affects (and even kills) fish and other aquatic life; because they cause annoying odors; because they impair domestic and livestock water supplies by affecting taste, odors, and colors and because they may give rise to scum and solids that make water unfit for recreational use.

Although some inorganic substances occur in this category, most oxygen demanding wastes are organic compounds. Pollutants in this category typically come from such sources as sewage, both domestic and animal; industrial wastes from food processing plants; wastes from paper mill activities; tanning operation by products; and effluent from slaughterhouses and meat packing plants.

Most compounds involved in this type of pollution have carbon as their most abundant element. One reaction they undergo, with bacterial help, is the oxidation of carbon to CO_2.

$$C+O_2 \rightarrow CO_2$$

As oxygen-demanding wastes rapidly deplete the DO of water, it has been important to be able to estimate the amount of these pollutants in a given body of water. The biochemical oxygen demand (BOD) of water has been a quantity related to the amount of wastes present. In a water sample, the BOD indicates the amount of dissolved oxygen used up during the oxidation of oxygen demanding wastes. It could be found out by incubating a sample of water for five days at 20°C. The amount of oxygen consumed (BOD) could be found out by chemical determination of the DO concentration of the water before and after incubation.

A BOD of 1 ppm has been characteristic of nearly pure water. Water has been regarded fairly pure with a BOD of 3 ppm, and of doubtful purity when the BOD value reaches 4 ppm.

The disappearance of plant and animal life is an obvious result of the oxygen depletion of water. This occurs by a direct killing effect or because of migrations to other areas.

The products of decomposition following these different pathways have been quite different, as shown in Table 3.1.

Methane (CH_4) has been odorless and flammable; amines are having a fishy smell; hydrogen sulfide is bad smelling and toxic; and some phosphorus compounds have unpleasant odors. When

TABLE 3.1

Comparison of Decomposition and Products Under Differing Conditions

Aerobic conditions	*Anaerobic conditions*
$C \rightarrow CO_2$	$C \rightarrow CH_4$
$N \rightarrow NH_3 + HNO_3$	$N \rightarrow NH$ + amines
$S \rightarrow H_2SO_4$	$S \rightarrow H_2S$
$P \rightarrow H_3PO_4$	$P \rightarrow PH_3$ and phosphorus compounds

these contributions are added to the odor of decaying fish or algae, it becomes apparent that a shift from aerobic to anaerobic conditions of decomposition is not which is favored by users of fresh air.

2. Disease-causing Agents

Water has been a potential carrier of pathogenic micro-organisms and can endanger health and life. The pathogens most frequently transmitted through water have been those responsible for infectious of the intestinal tract (typhoid and para-typhoid fevers, dysentary and cholera) and those responsible for polio and infections hepatitis. Historically, the prevention of water-borne diseases was the primary reason for pollution control in water. Modern disinfection techniques have greatly reduced this danger. This is not true for some large parts of the world, where for example, cholera epidemics are still common. The fact that such disease causing agents are under control must not result in a sense of false security. The occurrence of a polluted water supply leading to an outbreak disease is always a possibility. The responsible organisms are present in the feces or urine infected people and are ultimately discharged into a water supply.

Waste water released from municipalities, sanitoria, tanning and slaughtering plants, and boats may be sources of bacteria or other micro organisms which are capable of producing disease in men and animals, including livestock. Any fair-sized community at any given time has been likely to have some persons who are diseased so that disease micro organisms are almost always present in sewage.

There have been several types of human infections, not all of which are transmissible through water (see Table 3.2). Many of the diseases whose epidemics recurrently decimate human populations get transmitted by water, however, cholera and typhoid being important examples.

TABLE 3.2

Examples of Human Infections

A. Animal infections that are of public health importance because they are transmissible to man.

1. Tetanus from horses and cattle transmitted by inoculation or contract with animal faces.
2. Bubonic plague from wild rodents by insect (flea) bite.
3. Anthrax from herbivorous animals by direct contact.
4. Radies from dogs, bats, etc., by bites.
5. Bovine tuberculosis from cattle through ingestion or air-borne transmission.
6. Jungle yellow fever from monkeys through mosquito bites.
7. Several types of encephalitis from birds and fowl through mosquito bites.
8. Trichinosis from swine through ingestion.

B. Primarily human infections in which the infective agent has a certain period of extrahuman residence before transmission.

1. Schistosomiasis ("snail fever") from water from snails.
2. Urban yellow fever from mosquitoes.
3. Hookworm from soil by skin penetration.
4. Malaria (also a mosquito infection) from mosquitoes.
5. Typhus from lice.

(*Cond.*)

C. Infections that persist or multiply in the external environment and are transmissible from man to man.

1. Cholera, typhoid fever, bacillary dysentery, poliomyelitis, and infectious hepatitis from water and food through ingestion.
2. Staphylococcal and streptococcal diseases from food, air, and the proximate environment through contact and inhalation.
3. Smallpox from air, dust, and the proximate environment through inhalation.
4. Coxsackie and ECHO virus diseases from water through ingestion.

The identification of pathogens in water needs very large samples and many sophisticated techniques and is too time-consuming and expensive for routine pollution tests. The standard method involves determination of the most probable number (MPN) of coliform organisms in the water sample.

3. Plant Nutrients

Nutrients are an important limiting factor in the growth of all plants. With all other factors equal, the rate and profuseness of plant growth are proportional to the amount of nutrient available.

Plant nutrients (phosphorus and nitrogen) enter fresh and marine system and lead to or intensify eutrophication of these systems. These nutrients tend to accumulate in groundwater since it is out of the euphotic zone. As the groundwater moves laterally and reaches the surface waters, these materials add to the nutrient levels already present.

Plant nutrients like nitrogen and phosphorus are able to stimulate the growth of aquatic plants, which get interfere with water uses and later decay to produce disagreeable odors and add to the BOD of the water. Excess algal growths all of particular concern because they cause low dissolved oxygen levels and create treatment problems for municipalities and industries, in addition to interfering with recreational uses.

The enrichment of water with nutrients is a naturally occurring biological process called eutrophication. The term comes from two Greek words meaning "well nourished". This enrichment leads to other slow processes collectively referred to as natural aging of lakes. In some pollution articles, the term eutrophication includes both nutrient enrichment and lake-aging processes. The existence of peat and muck soils and deposits of coal and oil are proof that eutrophication and aging have been taken place in the past. The steps in eutrophication and aging of a lake have been as follows :

1. Streams from a drainage basin gradually bring soil and nutrients to a newly formed lake, increasing the fertility of the lake water.
2. The increased fertility leads to an accumulating growth of acquatic organisms, both plant and animal.
3. As living matter increasing and organic deposits pile up on the bottom of the lake, it becomes more shallow, warmer, and richer in nutrients.
4. Plants take root at the bottom and gradually occupy more and more of the space. Their remains accelerate the filling of the basin.
5. The lake gradually becomes a marsh and finally a field or forest as it has been overrun by vegetation.

The time needed for this aging process to be completed could be measured in thousands of years. The actual time depends upon the size and mineral content of the basin, and the climate.

The growth of green plants, including algae, needs the availability from fifteen to twenty elements which, together with their chemical symbols, have been given in Table 3.3. The relative amounts of these elements needed for plant growth depend upon the species involved. The amounts needed for the growth of three typical nuisance algae are given in Table 3.3. The values have been based on chemical analyses of the dried algae.

TABLE 3.3

Elements Necessary for Plant Growth

Carbon (C)	Magnesium (Mg)	Boron (B)
Hydrogen (H)	Calcium (Ca)	Vanadium (V)
Oxygen (O)	Sodium (Na)	Chlorine (Cl)
Nitrogen (N)	Iron (Fe)	Molybdenum (Mo)
Phosphorus P)	Manganese (Mn)	Cobalt (Co)
Sulfur (S)	Copper (Cu)	Silicon (Si)
Potassium (K)	Zinc (Zn)	

TABLE 3.4

Chemical Composition of Three Nuisance Algae

	Algae		
Element	*Microcystis*	*Anabena*	*Cladophora*
C	46.5 wt%	4[illegible].7 wt%	35.3 wt%
N	8.1	9.4	2.3
P	0.7	0.77	0.5[illegible]
K	0.8	1.2	6.1
Ca	0.53	0.36	1.7
S	0.27	0.53	1.6
Fe	0.27	0.08	0.23
Mg	0.17	0.42	0.23
Na	0.04	0.18	0.18
Mn	0.03	0.008	0.10
Zn	0.005	0.000	0.001
Cu	0.004	0.007	0.019
B	0.0004		0.0085

Most elements essential for plant growth have been available to the plant in amounts well in excess of the plant's needs. A few, however, have been present in amounts very close to the quantities required for plant growth. These may be used by the plant almost to the point of exhaustion. Considerable interest gets centered around these elements because they appear to behave as natural controls in preventing excessive plant growth. The growth of a plant will cease when the least available element gets depleted. Man's role in eutrophication appears to be one of increasing the amounts of these growth-limiting elements. Great changes in the availability of some elements in water have taken place through man's activities.

Both nitrogen and phosphorus occur in small amounts in natural waters, but their concentrations get greatly increased by the activities of man. As much as 80% of the nitrogen and 75% of the phosphorus added to surface waters originate from man-made sources. Table 3.5 gives estimates of possible nutrient

TABLE 3.5

Example of Possible Nutrient Discharges to Estuaries from Various Sources (Metric Tons Per Year)

	Nitrogen (N)	*Phosphorus (P)*
Sewage : population of 10 million, per capita generation 400 litres/day	6,000	15,000
River water : discharge 500 m^3/sec (approx. 15,000 cfs)	300	7
Subsurface seawater : 20 volumes of seawater mixing with each volume of river water	6,000	900
Storm-water runoff : 75 cm (30 inches) per year area	6,000	800

discharge (as P and N) into estuaries from a hypothetical community The community has been made up of 10 million people and the nutrients are put into a river discharging 15,000 cubic feet of water per second into the estuaries. It is regarded that 20 volumes of seawater dilute each volume of discharged river water.

The dominant source of phosphorus has been seen to be sewage, whereas nitrogen in various forms has been supplied in nearly equal amounts by sewage treatment plants, storm-water run-off, and subsurface seawater.

The phosphorus content of domestic sewage is becoming a topic of concern. It is estimated that up to 70% of this comes from the use of household detergents, and it should be subject to rather easy control procedures. The elimination of phosphorus compounds from detergents has been discussed at great length at another place.

Algae growth needs many different nutrients like carbon dioxide, nitrogen, phosphorus, iron, manganese, boron, cobalt, vitamins, hormones, etc. The elimination of any one of the essential nutrients would not allow algae growth but there is an active scientific controversy about which nutrients should be controlled. It is generally felt that control of nitrogen and phosphorus would be best but some scientists think that the availability of carbon is a much more important limitation, and the large amounts of phosphorus present in the sediments in lakes and rivers yield a vast reservoir available for the growth of algae.

4. Organic Chemicals

The exotic organic chemicals include surfactants in detergents, pesticides, various industrial products, and the decomposition products of other organic compounds. Analysis of polluted waters reveal the presence of a wide variety of these compounds. Some of these compounds have been found to be toxic to fish at very low concentrations, such as 1 ppm phenol. Many are not biologically degradable, or are degraded only very slowly.

Laundry detergents, a common constituent of waste water, have been another organic contaminant. Detergents consist of two major components : a surfactant or sudsing agent and a series of builders. The surfactant or sudsing agent can become a problem as it is not readily broken down by bacterial action and has a long residence time. Even in relatively low concentrations, it can cause water to form foam.

Most detergents manufactured in the early 19 0s were composed of approximately 40% sodium tripclyphosphate, the "builder" component. It is calculated that such phosphorus-bearing builders account for one half the total phosphate in most wastewaters. Due to this extremely high phosphorus concentration, these detergents are a major source of eutrophication in freshwater systems. In addition as detergents have been contributing a large amount of phosphorus to wastewater without a corresponding addition of nitrogen/phosphorus imbalance will take place when this material enters either marine or fresh-water systems.

Hydrocarbons, in the form of gasoline and motor oil, although insoluble in water, have been carried from roadways and parking areas in rainwater runoff.

In addition to the substances mentioned above, analysis of water reveals the presence of a wide variety of other organic chemicals. Some of these compounds are knawn to be toxic at very low concentrations, while many are not biodegradable or break down very slowly.

5. Oil

The production, distribution, and use of such large yearly quantities results in some oil contamination of the environment. Some of this contamination is accidental, some is not.

6. Inorganic Chemicals and Mineral Substances

This category of water pollutants have been including inorganic salts, mineral acids, and finely divided metals or metal compounds. These suostances enter natural waters because of activities in various smelting, metallurgical, and chemical industries; mine drainage and various natural processes. The presence of these brings about three general effects : the acidity, salinity, and toxicity of the water may be increased.

Inorganic chemicals of many types enter water from municipal and industrial waste waters and urban runoff. These pollutants are able to kill or injure fish and other aquatic life and they can interfere with the suitability or water for drinking or industrial use. Many of these inorganics are not only toxic but tend to concentrate in food chains.

A prominent example has been the occurrence of mercury in water. A number of industrial processes make use of mercury, some of which is eventually disposed of in waste-water effluents. It is now known, that anaerobic bacteria in bottom muds can convert inorganic mercury into methyl mercury (CH_3Hg^+), which can be concentrated in living things and lead to mercury posioning.

One potential pollutant arises in petroleum drilling, where bring gets discharged along with crude oil when the latter is pumped to the surface. In some places the brines have proved valuable sources of important minerals and elements, such as bromine, iodine and magnesium.

Another very important problem, is acid mine drainage. On exposed coal mine surfaces, minerals, containing sulphur (most notably iron pyrite, EeS_2) come into contact with air and water, forming sulfuric acid that has been carried into streams by waters draining from the mines. This takes place from abandoned mines as well as operating mines and has been most pronouuced in bituminous coal mines.

Acid mine drainage has been the primary source of pollutants that increase water acidity. This drainage adversely affects thousands of miles of streams and has been one of the most significant causes of water quality degradation in coal-producing areas. The actual pollutants present in mine drainage have been sulfuric acid (H_2SO_4) and soluble compounds of iron. These substances are formed because of a reaction between air, water, and pyrite (FeS_2) present in coal seams. Certain types of bacteria have been involved in the reaction, but their role has been not completely understood. This reaction can occur in both underground and surface mines.

During mining operations, in deep mines the strata between the coal seam and the surface have been invariably disturbed. Fissures appear through which water drains into the mine from many surface areas. This water containing the harmful pollutants is eventually discharged into surface streams, either naturally or through man-made processes. Damaging mine drainage is formed in surface mines if the surface run-off water comes in contact with pyrite-containing coal.

About 60% of the mine drainage pollution problems originate in mines once worked but now abandoned.

Large amounts of acid have been used to clean oxides and grease from metals. The used acids from these pickling operations have been becoming serious problems in some localized areas.

Due to the presence in solution of carbonate (CO_3^{2-}) and bicarbonate (HCO_3^-) ions, the pH of most productive, fresh natural waters lies between 6.5 and 8.5. This natural buffer system makes it possible for small amounts of acid or base to dissolve in the waters without making appreciable changes in the pH. These ions also represent an indispensable reservoir for carbon needed by aquatic plants during photosynthesis. Such use of these ions by plants can influence the buffering capacity of water, as there is a decided limit on the rate at which CO_2 (the source of CO_3^{2-} and HCO_3^-) can be obtained from the atmosphere to replace that used by plants.

Large influxes of strong acid have been able to overwhelm this buffering ability of water and bring about drastic drops in pH values. The effects of these changes depend upon the magnitude of the pH drop involved. Some effects have been as follows :

(*a*) *Aquatic life destruction.* At pH levels below 4.0, all vertebrates, most invertebrates, and many microorganisms get destroyed. Most higher plants are eliminated, leaving only a few algae and bacteria. Acid mine drainage is one of the primary causes of fish kills. Excessive precipitation increases the mine drainage output and compounds the problem. During high water seasons (winter and spring), the pH of the waters in some areas has been reported to be as low as 2.5.

(*b*) *Corrosion.* Water having a pH lower than 6.0 can bring about excessive corrosion of plumbing systems, boats, piers, and related structures.

(*c*) *Agricultural crop damage.* Acidity and alkalinity of irrigation water have been usually of little consequence over a pH range of 4.5 – 9.0, as the soil has been a buffered system. Problems may arise if the pH drops below 4.5. Such acidic water increases the solubility of such substances as Fe, Al and Mg salts. These ions, at the resulting high concentrations, have been sometimes toxic to plants.

Large amounts of salinity in water cause problems other than those related to human consumption. Dissolved inorganic and mineral substances exert abverse effects on aquatic animal and plant life and cause many irrigation problems in the agricultural industry. Damage to aquatic life is primarily related to the osmosis process, assuming the dissolved substances are nontoxic. Generally, the concentration of dissolved materials in body fluids is the maximum that an aquatic organism can tolerate. When these organisms are in contact with water containing higher concentrations, there is a tendency for water to move out of the cells of the organism into the surrounding water. The resulting increase in concentration within the cells of the organism can lead to death. Many fresh water species disappear when waters become brackish.

One of the most serious long term effects of increased salinity of water involves the use and re-use of water in irrigation. It has been estimated that about 25% of the irrigated land of the world is now affected to some degree by water salinity. Irrigation water brought onto a field always contains some dissolved salts. The concentration of these salts is in the range of 25 8000 mg/*l*. Plants extract water from the irrigated field, but most of the dissolved salts are excluded by the roots. Water that evaporates from the soil surface leaves dissolved salts behind. These two processes make residual salts to accumulate in the soil.

The toxic properties of numerous inorganic compounds, especially those of some of the heavier metallic elements, have been known for years. Some of these compounds are having desirable properties and so are routinely manufactured. Appropriate precautions are taken to insure the safety of individuals involved in the manufacturing process. The use of these compounds give rise to their introduction into the environment either directly or indirectly, intentionally or unintentionally. The recent detection of these metallic substances in air and water in concentrations approaching toxic levels has created a great amount of concern. The most toxic, persistent, and abundant of these compounds in the environment appear to be those of the metals mercury (Hg), lead (Pb), arsenic (As), cadmium (Cd), chromium (Cr), and nickel (Ni). These metals accumulate in the bodies of organisms, remain for long periods of time, and behave as cumulative poisons. More research is needed to determine the seriousness of these potential hazards.

7. Sediments

Sediments are soil and mineral particles which are washed from the land by storms and floodwaters, from croplands, unprotected forest soils, overgrazed pastures, strip mines, roads, and bullozed urban areas. Sediments are able to fill stream channels and reservoirs : erode power turbines and pumping equipment; reduce the amount of sunlight available to green aquatic plants; plug water filters; and blanket fish nests, spawn, and food supplies, thus reducing the fish and shelfish populations.

Sediments are almost regarded as a type of pollution due of the naturally occurring process of erosion. Sediments produced by that process do represent the most extensive pollutants of surface waters. It is estimated that suspended solid loadings reaching natural waters have been at least 700 times as large as the solid loadings from sewage discharge.

Sediment production is a tedious problem in some parts to the world. This variation is not all caused by man's activities, but also involves the type of soils, geology, topography, precipitation, and vegetation cover.

The detrimental effects of sediment in water are :

(*a*) *Stream channels, harbors, and reservoirs are filled.* This makes channels to overflow more easily, changes flow rates and depths of channels, and reduces the useful life of reservoirs. Expensive dredging is needed to counteract these effects.

(*b*) *Destroys aquatic animals.* Sediment settling on the bottom lowers fish and shelfish populations by blanketing fish nests and food supplies.

(*c*) *Reduces light penetration into water.* The reduction in the amount of sunlight penetrating into the water lowers the rate of photosynthesis by plants, which in turn results in a decreased production of oxygen needed for normal stream balance.

(*d*) *Water is clouded.* It increases the cost of treating water used for culinary purposes. Sediment passing through power plant turbines makes serious abrasion and wear. The hunting ability of fish get curtailed if they depend on their vision.

8. Radioactive Materials

Harmful radiotion may result in water environments from the wastes of uranium and thorium mining and refining, from nuclear power plants; and from industrial, medical, and scientific utilization of radioactive materials.

Many radioactive substances are lethal at relatively low concentrations and in minute amounts may be mutagenic.

Perhaps the greatest problem of radioactive pollution, as a result of uranium production, has been caused by the large quantities of "uranium tailings" produced. This finely divided solid material remains after useful materials have been leached out. Huge piles of these tailings are found in uranium producing areas.

These tailings create a radiation pollution problem as they contain radioactive decay products of uranium. Two of these radioactive materials are thorium-230 ($^{230}_{90}Th$) and radium-226 -($^{226}_{88}Ra$). Substances like these can be dissolved or eroded from piles of tailings by rainfall, and thus they can get mixed up into the general water supply. Radium and thorium is chemically similar to calcium and so get absorbed by the bones when taken into the body. Some waters in the Colorado River Basin have had the concentration of $^{226}_{88}Ra$ increase to about double the maximum level permissible for human consumption.

An important source of radioisotopes is nuclear weapons testing. The amount and variety of radioactive materials formed depends on the type of weapon tested. Some of the radio-isotopes have a very short half life and last only a few seconds or minutes, and some others may have a half-life of several hunred years. These materials reach the earth as radioactive fallout. Before the first test explosions were conducted, it was thought that fallout (dust and debris) would fall to earth quickly and at no great distance from the detonation site, thus the spread of radioactivity would be limited. However these assumptions have proved wrong, especially in the case of large explosions. Radioactive particles can, remain suspended in the air long enough to circle the globe many times, dispersing as they go. It is true, however, that the main effects of atmospheric testing are found in the immediate vicinity of the detonation site. Because of concern about such fallout in 1963, of the Limited Nuclear Test Ban Treaty was signed. According to the U.S. and U.S.S R. will limit testing of nuclear weapons to underground detonations.

The purpose of underground detonations, carried out in the U.S. at the Nevada test site, is to limit the amount of atmospheric fallout, and in majority of cases this is achieved. Occasionally accidental releases of radiation reach the atmosphere due to "venting" at the explosion site. When this happens, radioactive gases are forced up through the hole used to place the explosives. The leakage may take place through channels left the control and monitoring cables, natural cracks or fissures in the rock, fractures formed by the blast, or combinations of these. Sometimes venting occurs when the materials used to fill the hole are simply blown out. Leakage may immediately take place after the explosion or later, when rock and earth fall into the cavity formed by the explosion and leave a channel for gas venting.

In only 12 of 150 tests conducted at the Nevada site from 1961 through 1969 ventings were reported different amount of radiation are released. It is released near ground level and so poses no world-wide fallout problems, but it may result in very serious local hazards.

Atmospheric fallout, irrespective of the source, can have far-reaching effects and can be conveyed to man in a number of ways as shown in Fig. 3.1. The effects of radioisotopes on man can be illustrated. Strontium-90 ($^{90}_{38}Sr$), a component of radioactive fallout which has a half-life of 28 years, is chemically similar to calcium. Calcium is absorbed from the soil by plants and passed on to animals, where it is used in the formation of bones and teeth. Man gets calcium from both plant and animal sources like milk, vegetables, and cereal grains. Strontium is produced it from these sources and, due to chemical similarities to calcium it is also deposited in bones and teeth. The marrow of bones is the main site of blood-cell formation. The presence of radioactive strontium-90 in the surrounding bone tissue seriously curtails this which may result in production anemia or more serious disorders.

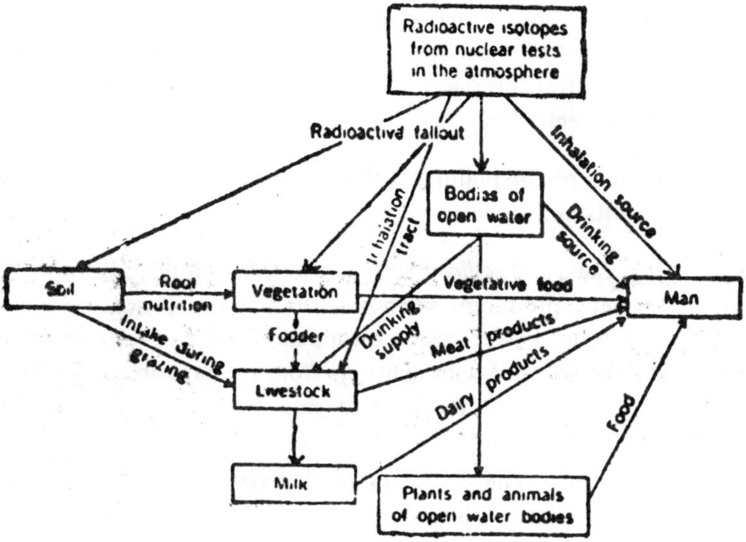

Fig. 3 1. Transmission of radioactive fallout to man.

Cesium-137 ($^{137}_{55}Cs$) which is chemically similar to potassium is a common constituent of all living cells. Cesium from fallout passes to man through contaminated meat and dairy products or contaminated grains and leafy vegetables. Cesium contamination of foods results in the same effects as strontium contamination The soft parts of the body, specially the muscles, are badly affected by the presence of cesium.

More and more of nuclear energy is being produced so as to meet the ever increasing demand for electrical power. Nuclear generators are better as compared to more commonly used types. Because, nuclear energy is cheaper to use than that from conventional sources such as fossil fuels or falling water. Moreover, nuclear generators do not produce any SO_2 or particulate pollutants commonly produced by fossil-fuel burning generators. By the year 2000, nuclear power is expected to be the main source of electric power generation.

Nuclear power plants are different from conventional electricity generating facilities. In the conventionally fueled power plant, fossil fuel is burnt to produce heat while in a nuclear plant the heat is generated by a nuclear reaction. In both the heat produces steam which turns a generator and produces electricity.

Four types of pollutants that are formed by nuclear generating plants are as under:

(*a*) *Low level radioactive liquid wastes.* Radioactive isotopes are produced when impurities in the primary coolant water and corrosion products from coolant pipes are bombarded with neutrons from the core area. This can be avoided to some degree by using demineralized coolant water.

(*b*) *Heat.* The secondary cooling system takes away huge amounts of heat from the reactors and mixes it into natural water supplies.

A variety of research laboratories, including medical and biochemical facilities, release radioactive wastes into the air and sewer systems. It is not now feasible to collect and dispose of these materials. Nor is it feasible to collect and bury the large amounts of radioisotope-containing liquids, solids, and gases forward by laboratory uses other than research, like diagnostics or treatment. Presently, the amounts of radioactive waste discharged from such laboratory sources poses no threat to natural communities of plants, animals, and man.

(c) *Liquid and gaseous wastes from fuel elements.* Complete sealing of the fuel in steel or zirconium containers is apparently impossible to attain or sustain. Minute cracks allow fission products to escape into the primary coolant. This further complicates the disposal problem of low level wastes discussed earlier.

(d) *Fission products.* Within 1·3 years, fission products (the ashes of nuclear fuels) accumulate to the point that they absorb sufficient neutrons to slow or stop the chain reaction. Then the extremely radioactive fuel elements are removed and shipped in special containers to a fuel reprocessing plant. In these plants the fission products are separated from the remaining usable fuel. The fuel is returned to the reactor where it is used and the waste fission products are stored, in liquid form, in huge underground stainless steel tanks. Hundreds of millions of gallons of these high level wastes are now in storage.

9. Heat (Thermal Pollution)

Vast amounts of water find use for cooling purposes by steamelectric power plants (and other industries to a lesser extent). Cooling water is discharged at a raised temperature, and some rivers may have their temperatures so high (even up to 40°) that fish life completely gets eliminated and the river becomes useless for assimilation or pollution further cooling.

Increasing the water temperature of a system is harmful since it generally alters the chemical, physical and biological characteristics of that system. In addition to the possibility of decreasing or eliminating various aquatic forms, it may also stimulats spawning at a time of year when food supplies are limited. This leads to starvation of the newly spawned individuals in the population. High temperatures also decrease the density and viscosity of water, causing an increased settling rate of suspended solids. Evaporation rate is increased and, in marine systems, this may lead to localized areas of abnormally high salinity.

Heat is not ordinarily thought of as a pollutant by many people, at least not in the same sense as a corrosive chemical. However, the addition of excess heat to a body of water brings about adverse effects as numerous as many of the chemical pollutants. This serious problem of thermal pollution originates primarily with the practice of using water as a coolent in many industrial processes. Most water used for this purpose gets returned, with the added heat, to the original sources. At present, about 70% of the water diverted, to industrial use serves as a cooling medium.

Used coolent water frequently may be having a temperature 20°F higher than the river or stream to which it gets returned. This is added heat raises the temperature of the natural waters, with the results that : (1) the amount of dissolved oxygen in the water gets decreased; (2) the rates of chemical reactions have been increased; (3) false temperature cues have been given to aquatic life; and (4) lethal temperature limits may be exceeded.

The decreasing ability of water to contain dissolved oxygen as the temperature increases is shown graphically in Fig. 3.2.

The addition of heated water to a cooler, body of water may accelerate the lowering of DO levels due to density differences between the two. The less dense warm water tend to form a layer on top of the cooler, more dense water. This occurs particularly when the body of cool receiving water is deep. The resulting blanket of "hot" water cannot dissolve as much atmospheric oxygen as the underlying cold water, which is denied contact with the atmosphere. Normal biological reduction of the DO level of the atmospherically unreplenished lower layer may give rise to anaerobic conditions.

Another effect of this stratification may show up downstream from a dam when the oxygen deficient lower level get discharged through the lower gates of a dam. Serious effects on downstream fish life may result. Also, the ability of the stream below the dam to assimilate oxygen demanding wastes will get curtailed.

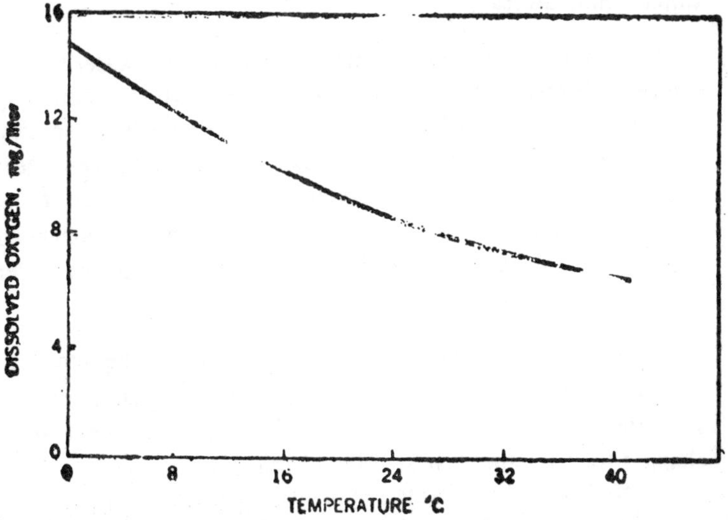

Fig. 3.2. The quantity of oxygen dissolved in water is related to temperature. The higher the temperature, the lower the oxygen content of the water. Thermal pollution can lower the oxygen content below the point necessary to sustain many animals.

In thermally polluted water, fish need more oxygen due to an increased respiration rate. However, the available oxygen in such water gets decreased. Thus, thermal pollution affects fish in a double-barreled fashion.

Other reactions are also influenced. Trout eggs hatch in 165 days when incubated at 37°F. When water temperatures have been 54°F, only 32 days are needed and no hatching takes place at water temperatures in excess of 59°F. Such a result can be disastrous to fish populations.

The life cycle and natural processes of many aquatic organisms have been closely and delicately geared to water temperature. Fish often migrate, spawn, and are otherwise distributed in response to water temperature cues. Shellfish, such as oysters, spawn within a few hours after their environment reaches a critical temperature. These normal life patterns of aquatic organisms can get completely disrupted by artificial changes in water temperatures.

Fish have been found to vary widely in their water temperature preferences. An indication of this has been the commonly used classification of fish as either cold- or warm-water species. An upper lethal, temperature limit exists for each species. It is known that temperatures well below the lethal value can bring about stress in organisms; hence the much lower recommended values. Lethal limits can be exceeded, in some examples under conditions of warm water waste discharge.

Another factor to be considered in addition to the temperature change has been the rapidity with which the change occurs. Fish are able to acclimate themselves to moderate temperature changes (below lethal levels) if the change has been not a sudden one. Normally a moderate temperature change over a 30-40 hour period could be handled by the fish. For instance, 95% of the eggs of largemouth bass perish when suddenly transferred from water at 6S-70°F into water at 85°F. However, if similar eggs are kept in water and the temperature gradually increased to 85°F over a 30-40 hour time period, 80% of the eggs survive.

An additional threat to aquatic life has been created by the common practice of chlorinating coolant water prior to use. This is carried out to prevent bacterial growths that clog pipes. The chlorine affects organisms in the area receiving the used water until dilution occurs. The chlorine has been able to kill microorganisms important in some food chains and may in this way exert a negative influence on fish populations.

A problem of thermal population not related to aquatic life has been the reduced cooling capability of warmed waters. It is important because nuclear reactor power sources need about 50% more cooling water for a given temperature increase than similar power plants using fossil fuels, and it appears that nuclear plants will find use extensively in the future.

Different solutions to the thermal pollution problem have been postulated and some have been already in use. Several industries have incorporated cooling towers into their operations to remove heat from cooling water before returning it to the natural water supply. Two types of towers, wet and dry, have been used. Water is run over baffles in a thin layer in the wet tower. Cool air, entering at the bottom, circulates upward and removes heat from the water. In dry towers, air is forced over water-containing pipes by huge fans. Heat gets exchanged by radiation and convection from the pipes

The use of cooling ponds or lakes has been another alternative. These ponds could act as ice-free wintering areas for waterfowl in northern locations or as a means of extending the range (northward) of certain fish. Both of these have been positive results of thermal pollution.

SOURCES OF CONTAMINATION OF WATER POLLUTION

Introduction

The major sources of water contamination have been domestic, industrial and agricultural waste, as well as solid waste thermal pollution, shipping water pollution and radioactive waste.

Domestic Water Pollution

It includes waste water from homes and commercial establishments. Domestic waste water arises from many small sources spread over a fairly wide area but is transmitted by sewers to a municipal waste treatment plant.

Generally, the impurities in domestic wastes get diluted and seldom total more than 0.1% of the total mass. This material has been largely organic and gets oxidized by bacterial decomposition to nitrate, phosphate, carbon dioxide, and water. As this type of decomposition needs the use of dissolved oxygen, it place an oxygen demand on the system. Because of this tendency to remove oxygen in the decomposition process, a common indicator used to monitor this type of input in receiving water has been the BOD (Biological Oxygen Demand) Test. In this analysis the amount of oxygen for decomposition is measured over a 5-day period. In systems receiving significant amount of organic material, the bacterial decomposition will remove large amounts of dissolved oxygen, which causes oxygen depletion. These systems are considered to have a high BOD. Conversely, when the input of contaminants in absent or minimal, the oxygen levels will not be drastically reduced because the bacteria will not need an excessive amount of oxygen to decompose small amounts of materials. In these cases there would be a low BOD on the system.

Industrial Water Pollution

This occurs in large amounts in specific locations, making collection and treatment fairly simple to accomplish. There are water-using factories which are discharging wastes with a total BOD load about three to four times as large as the load from the sewered population. Only about 7 or 8% of industrial waste waters have been disposed of in municipal sewer system.

Wastes from textile manufacturing processess could be generated from the washing out of the impurities in the fibers, as well as in the discarding of chemicals used in the processing of the fibers. Generally, these wastes have been organic, have a high BOD, and are extremely alkaline.

Food-processing wastes from meat, dairy, and sugar-beet processing, as well as brewing, distilling and canning operations, generate large amounts of organic by-products that have been disposed of in waste-water. When the waste-water is discarded, along with these by-products, it leads to high BODs and a consequent oxygen depletion in the receiving water via the same bacterial process involved in the decomposition of domestic wastes.

The effluent released from pulp and paper processing operations has been a mixture of chemicals used in the digestion of raw wood chips, cellulose fibres, and dissolved lignin. This waste-water also has paper and wood preservatives, like pentachlorophenol and sodium pentachlorophenate, as well as methyl mercaptan, all of which have been toxic to fish. This effluent has been brownish in colour and lowers the photosynthetic rate of aquatic communities by hindring sunlight penetration into the water column. Consequently, the organic wastes from these plants increase the BOD of the receiving water, while colour imparted to the water interferes with sunlight penetration' reduces photosynthesis, and further lowers oxygen levels.

Metal industries keep a wide array of contaminants in their waste-water. The specific contaminants and concentrations depend solely on the particular manufacturing process employed. For example, steel mills employ and contaminate water in the coking of coal, the pickling of steel, and the washing of flue gases from blast furnaces. These waters, after use tend to be acidic and have various deleterious substances, like phenol, cyanogen, ore, coke, and fine suspended solids. Other industries release traces of the metals produced or plated in their waste-water. Metals commonly found in these waste-waters have been chromium, mercury, nickel, lead, copper, and cadmium.

A variety of contaminants enter marine and fresh water environments in the effluents release from the various chemical manufacturing plants. The release of acids results not only from acid manufacturing processes but from practically all other chemical manufacturing processes as well. In addition, synthetic fibres (*e.g.*, rayons), bases, pesticides, and other organic and inorganic chemicals have been added, depending on the products being manufactured.

Most of the Indian rivers and fresh-water streams are seriously polluted by industrial wastes or effluents (Table 3.6) which come along waste-waters of different industries such as petrochemical complexes; fertilizer factories; oil refineries; pulp, paper, textile, sugar and steel mills, tanneries' distilleries, coal

TABLE 3.6

Some Indian Rivers and their Major Sources of Pollution

Name of the River	*Sources of Pollution*
1. Kali at Meerut (U.P.)	Sugar mills; di leries; paints soap, rayon, silk, yarn, tine and glycerine industries.
2. Jamuna near Delhi	D D.T. factory, sewage, Indraprastha Power Station, Delhi.
3. Ganga at Kanpur	Jute, chemical, metal an‑l surgical industries; tanneries, textile mills and great bulk of domestic sewage of highly organic nature.
4. Gomti near Lucknow (U.P.)	Paper and pulp mills; sewage.
5. Dajora in Bareilly (U.P.)	Synthetic rubber factories.
6. Damodar between Bokaro and Panchet	Fertilizers, fly ash from steel mills, suspended coal particles from washeries, and thermal power station.
7. Hoogly near Calcutta	Power station; paper pulp, jute, textiles, ch‹mical mills, paint, varnishes, metal, steel, hydrogenated vegetable oils, rayon, and soap, match, shellac, and polythene industries and sewage.
8. Sone at Dalmianagar (Bihar)	Cement, puple and paper mills.

9. Bhadra (Karnataka)	Pulp, paper and steel industries.
10. Cooum, Adyar and (Bukinghum canal (Madras)	Domestic sewage, automobile workshops.
11. Cauvery (Tamil Nadu)	Sewage, tanneries, distilleries, paper and rayon mills.
12. Godavari	Paper mills.
13. Siwan (Bihar)	Paper, sulphur, cement, sugar mills.
14. Kulu (between Bombay and Kalyan)	Chemical factories, rayon mills and tanneries.
15. Suwao (in Balrampur)	Sugar industries.

washeries, synthetic material plants for drugs, fibres, rubber, plastics, etc. The industrial wastes of these industries and mills include metals (copper, zinc, lead, mercury, etc.), detergents, petroleum, acids, alkalies, phenols, carbamates, alcohols. cyanide, arsenic, chlorine and many other inorganic and organic toxicants. All of these chemicals of industrial wastes have been toxic to animals and may bring about death or sublethal pathology of the liver, kidneys, reproductive systems, respiratory systems, or nervous systems in both in vertebrate and vertebrate aquatic animals (Wilbur, 1969). Chlorine which is added to water to control growth of algae and bacteira in the cooling system of power station, may persist in streams to cause mortality of plankton and fish. Heavy fish mortility in river Sone near Dehrion-on-sone in Bihar is reported to cause by free chlorine content of the chemical wastes discharged by factories near Mirzapur in U.P.

Mercury like other heavy metals such as lead and cadmium has cropped up as a toxic agent of serious nature. Mercury, a by-product of the production of vinyl-chloride, is used in many chemical industries and it is also a by-product of some incinerators, power plants, laboratories and even hospitals, (Aaronson, 1971). In Japan, illness and even death occurred in the 1950s among fishermen who ingested fish, crabs, and shell-fish contaminated with methyl mercury from Japanese coastal industries. This mercury poisoning produced a crippling and often fatal disease called Minamata disease. Initial symptoms of minamata disease included numbness of the limbs, lips, and tongue, impairment of motor control, deafness, and blurring of vision. Cellular degeneration occurred in the cerebellum, mid-brain, and cerebral cortex and this led to spasticity, rigidity, stupor and coma. In Japan in 1953, due to Minamata disease 17 person died and 23 were become permanently disabled.

In India, all the 14 major rivers have become polluted. The river Damodar is perhaps the most heavily polluted river. River Mini-Mahi in Baroda has been other heavily polluted river which is having a variety of industrial and petrochemical wastes. The river Cooum flowing through Madras has been got polluted by sewage so much that not even the zooplanktons have been able to thrive in it. One litre of Cooum water is having as much as 900 mg or iron, 275 mg of lead, 1313 mg of nickel and 32 mg of zinc. Besides heavy metals, very high level of phosphates, slicates and nitrates also occur in the water. Sulphate levels ranging from 80-408 mg l^{-1} were the highest recorded among Indian rivers (The Times of India, May 27, 1987). The river Ganga from Hardwar to Calcutta is regarded as one unending sewer which is fit only to carry urban liquid waste, half burnt dead bodies, carrion, pesticides and insecticides. Nearly 312 industrial units are dumping their waste into the river, only a dozen have effluent treatment facilities. The 27 cities contribute 902 million litres of waste-water to the river each day. The water of Ganga affect the health of 250 million people of northern India.

Many of our lakes, notably the Dal lake, are becoming darkened, smelly and choked with excessive growth of algae.

Agricultural Water Pollution

It includes sediments, fertilizers, and farmanimal wastes. These pollutants can all enter waterways as runoff from agricultural lands but farm animal wastes are an especially large problem near the large feedlots on which thousands of animals are concentrated.

Agricultural waste includes the pesticides that are sprayed on crops, as well as sediment, fertilizers, and plant and animal debris that are carried into waterways during periods of rainfall or as runoff and during the irrigation of farmland. Wastes generated by farm animals are also included in this category. The practice of keeping large numbers of animals in a small area has led to an excess of animal wastes generated in and confined to a given area, where it has been economically impossible to distribute wastes for reuse as fertilizers. These materials become a problem when they are allowed to enter waterways during the cleaning of the confinement areas or during periods of heavy rainfall, when runoff carries them into adjacent waterways. Since these wastes are organic, they increase the BOD of the receiving waters.

Inorganic fertilizers, being plant nutrients, lead to over-fertilization of waterways when they enter these systems through runoff or during irrigation. The addition of excess plant nutrients can lead to a disturbance of the phosphorus/nitrogen balance in these systems as well as excessive plant growth. When the plants die, they settle to the bottom and, since they are organic, increase the BOD of the system during decomposition.

Soil erosion poses a fourfold problem. It increases the normal rate of filling of the waterways into which it washes, decreases the amount of fertile land for crop production, carries pesticide coated soil particles into the water, and decreases the transparency of the water, which limits photosynthesis. In addition, the sediment carried into fresh water systems tends to clog the gills of adult fish and settles out over incubating eggs, causing suffocation.

A remarkably large number of pesticides have come into widespread use in recent years. Many of these compounds have been not only non-biodegradable but are also only slightly soluble in water. Consequently, when sprayed on cropland they remain in the soil for long periods of time. During periods of heavy rainfall or when the crops are irrigated, they tend to be carried, as suspended particles, into surface, marine, or groundwater systems. In both fresh and marine systems they enter the food chain, undergo concentration in nontarget organisms, and increase in animal tissue to alarming levels. In surface, fresh, and groundwater systems they may also enter the drinking-water supplies of various communities.

Solid Waste Pollution

Solid waste varies in composition with the socioeconomic status of the generating community. The following materials could be classified as solid waste :

(a) Garbage, which includes all decomposable wastes from households, as well as from food, canning, freezing, and meat-processing operations that are not disposed of in wastewater.

(b) Rubbish includes all nondecomposable wastes. These materials may be either combustible or noncombustible. Combustible materials would include garden wastes, cloth, and paper. Non-combustible materials include masonry, some chemicals, metals, and glass.

(c) Sewage sludge is generated from the settling processes in primary, secondary, and tertiary treatment methods as well as the solids from cesspools, which must be removed periodically.

(d) Miscellaneous materials include industrial wastes, like chemicals, paint, and explosives, as well as mining wastes, such as slag heaps and mine tailings

The disposal of solid waste poses many problems, depending upon both the type of waste and the disposal method employed. The majority of the waste classified as combustible – rubbish, garbage. and sewage sludge--has been disposed of by one of three major methods : incineration, using it as landfill, or disposal by ocean dumping. Incineration generally gives to air pollution; landfill operations or ocean dumping lead to water contamination. If landfill disposal is used, the material, as it decomposes, will dissolve in or become suspended in the rain-water percolating into the ground and thus into subsurface aquifers. This tends to contaminate not only groundwater but also surface waters, since the two systems are ultimately inter-connected. Because of the highly organic nature of this material, a large BOD is placed on the receiving waters, and the sediment becomes coated with a highly organic ooze. Non-combustible materials are generally disposed of in landfill sites or by ocean dumping.

Thermal Pollution

Thermal pollution takes place because many electric-generating companies use water in the process of cooling their generators. This heated water is then released into the system from which it was drawn, causing a warming trend of the surface waters. Thermal pollution results when the heated effluent is released into poorly flushed systems. In these cases permanent temperature increases often result, which tend to decrease the solubility of dissolved oxygen. In lakes it also becomes possible to bring about nutrient redistributions and prolong summer stagnation periods.

When heated water gets released into large, well-flushed marine systems there is little if any permanent temperature rise. There are, however, problems related to the operation of plants utilizing marine waters in the cooling process. Evidence reveals that seawater tends to corrode the cooling pipes, which are generally constructed of a copper nickel alloy termed Monel. These metals readily dissolve in the heated seawater and are then released into the marine environment together with the heated effluent. This adds to the nickel and copper concentrations of these systems. In addition, the screens covering the water-intake pipes rapidly foul with marine organisms, which decreases the flow of water into the plant. The screens have been commonly cleaned by using a concentrated detergent solution or copper sulfate. These cleaning materials have been then released into and contaminated the surrounding waters.

Shipping Water Pollution

It includes both human sewage and other wastes, the most important of which has been oil.

Oil pollution results from accidents involving oil tankers and from spills at offshore oil drilling sites A more persistent source of oil pollution results from the practice of oil tankers, after they deliver the oil, to fill the empty tanks with seawater to act as ballast for the return trip. Prior to docking, the seawater ballast, contaminated with the oil that remains in the tanks, is discharged. Although this practice is illegal, it is difficult to prevent. The alternative is to pump the contaminated seawater into tanks at port. This is not only inconvenient but uneconomical. In addition, there will remain the problem of disposal when the disposal tanks are full.

Radioactive Waste Pollution

The major sources of radioactive wastes have been nuclear explosives, accidents at nuclear power plants, fuel-reprocessing plants, and research laboratories and hospitals that release these wastes into the atmosphere or into wastewater. Presently, most interest centers on radioactive iodine and strontium, since man is at the end of the food chains that concentrate these elements.

The significance of the various parameters discussed and their use as water quality indicators has been summarized in Table 3.7.

TABLE 3.7

Water-quality Indicators

Parameter	*Significance*	*Level*
1	2	3
Dissolved oxygen	General indicator of water quality; source of O_2 for respiration	Minimum acceptable level, 4-5 mg/liter; 10-15 mg/liter for reproduction of desirable fish
Total dissolved solids	Clog fish gills, bury eggs, reduce light penetration, increase heat absorption	Dependent on location
Total dissolved solids	Represents total mineral content which may or may not be toxic	A maximum of 400 mg/liter for diverse fish populations
BOD	Amount of dissolved oxygen removed during decomposition of organic matter in a given time; a general indicator of contamination due to biodegradable organics	BOD Water Status 1 mg/liter Very clean 2 mg/liter Clean 3 mg/liter Fairly clean 5 mg/liter Doubtful 10 mg/liter Contaminated
COD	Indicates the concentration of materials oxidizable by chemical reaction	0-5 mg/liter indicates very clean streams
pH	Indicates the addition of acids or bases	pH depends on actual system
Iron	Excessive amounts can clog fish gills; indicates	A maximum of 0.7 mg/liter for diverse fish

(*Contd*)

1	2	3
	drainage from iron-bearing sediments, mines, industrial processes	populations
Manganese	Concentration low in natural systems due to low solubility; high concentrations indicates contamination	A maximum of 1 mg/liter is a common criterion for stream quality
Copper	Indicates drainage from copper-bearing sediment, mides, plating, or other industrial sources	A maximum of 0.02-10 mg/liter is a common criterion for stream quality
Zinc	Indicates mine drainage or industrial input	A maximum of 1 mg/liter is a common criterion for stream quality
Hg, Cd, Pb, Ni, Cr, Ag, etc.	Indicates industrial input	A maximum of 1 mg/liter is a common criterion for stream quality
Nitrate	A major plant nutrient; in high-concentrations it can promote excessive plant growth; major sources are fertilizers, sludge, and sewage	A maximum of 0.3 mg/liter to prevent excessive fertilization of streams
Phosphate	A major plant nutrient; major sources are detergents, fertilizer, sewage	A maximum of 0.03 0.40 mg/liter total inorganic phosphate is a common criterion

Mercury

Mercury compounds, like many others, are dispersed throughout rocks, soil, air, water, and living organisms by a complex system of physicals chemical and biological processes.

Mercury has been used in a wide variety of ways. The largest user has been seen to be the chlor-alkali industry, which produces chlorine (Cl_2) and caustic soda ($NaOH$) by the electrolysis of salt ($NaCl$) solutions. The function of mercury in this process is to serve as the cathode of the electrolysis cell.

The second greatest consumption of mercury has been in the production of electrical apparatus. This application encompasses a large variety of uses.

The use of mercury and its compounds as fungicides constitutes the third largest category of mercury consumption. In this application, use is made of the toxicity of the substances toward livimg organisms. These substances have been used to destroy fungi in the paint, pulp and paper, and agricultural industries. Mercury compounds have been added to paint that is to be used in humid regions to act as a latex preservative and a mildewcide. Paints for use on ships must contain "anti fouling" agents to prevent barnacle growth. Mercury oxide (HgO) is often used in this application. Mildew proofing of marine paints has been also necessary, and phenyl mercuric acetate or its derivatives have been commonly used.

This last compound, phenyl mercuric acetate (PMA), has been the most important organomercury compound of commerce. The formula for this widely used substance has been as follows :

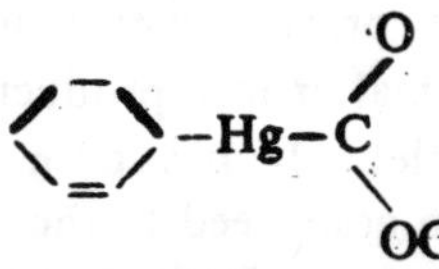

Other human activities contribute to mercury levels also. Sludge from sewage plants is having enough mercury, in some eares, to make it unsuitable for use as a soil fertilizer. The other sources are such commonly used materials as disinfectants, pharmaceuticals, and paints.

Mercury Amplification in Food Chain

Mercury is regarded to get into the food chain by two routes, water and seed dressings.

It is becoming evident that mercury is concentrated in living tissues. This concentration process has been involving food chains in which small organism containing mercury are ingested in quantity by larger organisms, which, in turn, are eaten by still larger one and so on. The organisms further up the chain obtain all the mercury collected by those down the chain. This amplification can given rise to high levels of mercury in the tissues of animals near the top of the chain. Man is at the top of every food chain in which he is involved.

In the case of fish, tiny marine diatoms are having mercury compounds adsorbed to their surfaces. Fish, upon eating these or other small animals or plants, concentrate the mercury in their tissues. Some mercury also gets absorbed by fish through their gills from surrounding water. Because of this process, the tissue of the fish has a higher mercury level than the surrounding water. For example, trout exposed to mercury levels of 0.1 ppm for 1 hour per day has been reported to the having a tissue level of 4 ppm of mercury after 10 days exposure.

Some evidence reveals that mercury from seed dressings can undergo translocation in plants and animals. Translocation in plants has been involving the transfer of mercury from seed to crop. In animals, the process makes mercury to move to various parts of the animal or into products obtained from the animal (eggs, for example). It has been possible for mercury to get translocated from grain seed to the harvested grain, then to a hen, and finally to eggs. Data exist to support this idea

Mercury Poisoning and Biological Methylation

The occurrance of mercury poisonings has been well documented fact. Widespread incidents have been reported only during the last two decades These incidents occurred due to the eating of mercury-contaminated fish or the eating of seed grain treated with mercury compounds.

Although the toxic action of mercury in the body has been not completely understood, recent research has revealed some information about the problem. The primary facts determined have been as follows :

1. All mercury compounds have been toxic to the body when present in sufficient quantity.
2. Different mercury compounds exhibit somewhat different characteristics of toxicity, distribution, accumutation, and retention times in the body.
3. Certain biological transformations may occur in the environment or the body which change mercury compounds of one type into other types.
4. The ultimate effects of mercury in the body appear to be inhibition of enzyme action and cellular damage caused by the ability of mercury to bind tightly with sulfur containing groups in molecules which are present in both enzymes and cell walls. Such binding is able to inactive enzymes and upsets chemical reactions which they catalyze in the body. The membrane properties of cell walls get disrupted by the binding of mercury, and normal cellular activities are inhibited.
5. The damage done to the body by mercury has been generally usually permanent. There has been no known effective treatment.

Inorganic mercury tends to accumulate in liver and kidney tissue. This causes damage, but also offers a means for rapid elimination through the urinary system. Metallic mercury vapor is having a great capacity for diffusion through the lungs into the blood and then into the brain, where serious central nervous system damage takes place. Normally, inorganic mercury, in the form of compounds, does not remain in the body long enough to accumulate to dangerous levels.

Upon introduction into the body, arylmercurials break down readily into inorganic mercury compounds and behave as indi cated above. The compounds PMA, behaves this way.

Alkylmercurials represent the greatest health hazard due to a number of reasons :

1. They easily penetrate into and accumulate in brain tissue. This is accomplished in part since they diffuse easily through biological membranes.
2. They exhibit long retention times in the body. This makes the mercury concentration to build up even though dosages may be low. The estimated half-life of these compounds in the body has been 70 days.
3. Alkylmercurials can be produced from inorganic mercury by the action of certain anaerobic micro-organisms. This transformation occurs readily in the mud of rivers and lakes. The extent to which it takes place in the human body is not known, but micro-organisms found in some animal intestines are able to accomplish the feat.

The classes of mercury and their relationships to each other have been given in Fig. 3.3.

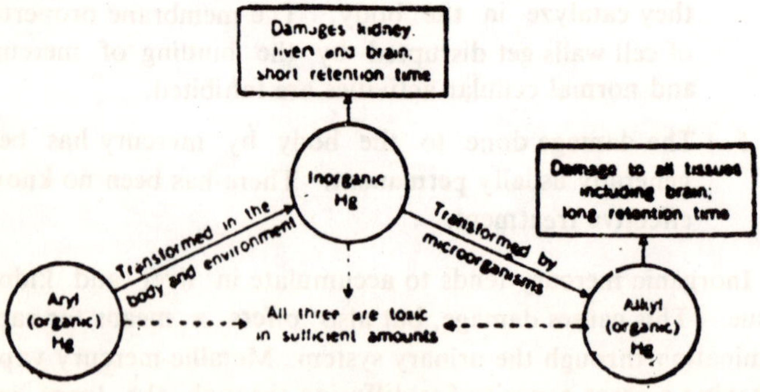

Fig. 3.3. Relationship between classes of toxic mercury.

Methylmercurials appear to be the most dangerous and toxic of the mercury compounds. They can enter the environment in several way – directly, indirectly. intentionally, or unintentionally. A direct and intentional route could be represented by the use of methylmercurials on seed. A direct but unintentional routes has been provided when methylmercurials used in industry or formed as a waste product of industrial processes are dumped into natural waters. The episode at Minamata Bay was caused in this way. An indirect route has been provided when other mercury compounds get methylated (transformed) by biological organisms.

A very serious example of the latter route has been involving inorganic and phenyl mercury from industrial sources. The metallic mercury or compound, upon settling into the bottom mud of lakes, gets converted by anaerobic microorganisms into CH_3Hg^+ or $(CH_3)_2Hg$. This latter compound has been very volatile and escapes from the mud into the water. The $(CH_3)_2Hg$ has been stable in alkaline solutions, but when exposed to acidic surroundings, it changes to CH_3Hg^+. This ion has been soluble in water and concentrates in living organisms, usually appearing in body lipids. This accumulation occurs due to direct absorption of the ion or an amplication by a food chain process.

In order to prevent environment pollution by mercury, following facts must be kept in mind :

1. The use of all alkyl mercury pesticides must be terminated.

2. The use of all other mercurial pesticides should be restricted to areas of demonstrated need.

3. All industrial users of mercury must be required to reduce discharges of mercury into the environment to levels approximating normal background levels.

The implementation of the above recommendations will not completely solve the problem. Sufficient mercury contamination has been present in the bottom muds of rivers and lakes to continue to yield highly toxic CH_3Hg^+ into the waters for years to come. Some sort of decontamination is needed. Large scale pilot experiments on decontamination methods have been postulated ;

1. Bottom sediments get covered with fresh, finely divided materials having high absorption capabilities.
2. Bottom sediments get buried under inorganic, inert materials.
3. The mercury-containing bottom sediments have to be removed by dredging or pumping.

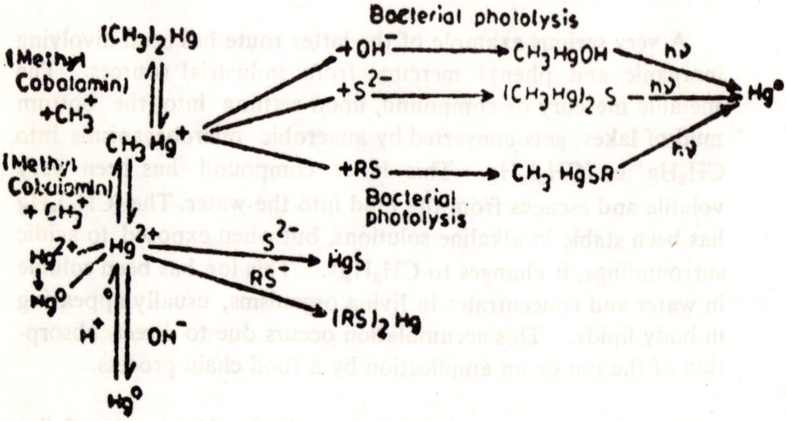

Fig. 3.4. Environmentals ehemistryof Hg.

Sources of Weter-borne Lead Pollution

Public water supplies in most countries rarely contain clinically significant quantities of lead. Most samples with high levels have been taken from waters located near metal working industries.

Some series outbreaks of lead poisoning occurred in Europe a number of years ago. The source of the lead was found to be lead pipe used in interior and exterior plumbing. The hard waters contain carbonate and sulphate ions which react with lead to form a water-insoluble protective coating of $PbCO_3$ and $PbSO_4$.

Two problems similar to the one of lead in water have recently become of interest. These problems involve moonshine whisky and glazed ceramic containers.

The consumption of illegally produced whiskey has been a cause of lead poisoning in some parts of the world. The source of this lead in whiskey is lead solder used in the tubing of distillation units, and lead-containing automobile radiators used as condensers.

Lead-containing ceramic glazes have been a serious source of lead poisoning when used on containers for foodstuffs. It has been found that highly acidic liquids such as apple juice may dissolve the glaze and release lead into the liquid if the glaze is not properly formulated and applied.

Soils in some areas have been still heavily contaminated with the stable compound lead arsenate, which was a widely used pesticide prior to World War II. Its use has been discontinued in favour of organic pesticides in many areas.

Another dietary problem has been possible among small children who are known to have a penchant for nibbling on materials not normally considered part of a well-rounded diet. In many countries lead is still used as an ingredient of interior house paint. Children living in pre-1940 structures have suffered lead poisoning from eating flaking paint. Children between the ages of 1 and 6 years are the main victims, and 85-90% of the cases involve youngsters in the 1-3 year age group.

The chemical form of lead has been an important factor which is affecting its biological behaviour in the body. Organic lead, compounds, such as tetraethyl lead, get readily absorbed by the body through the skin, or mucous membranes. This a serious problem only to a small group of occupationally exposed workers, such as those involved in the manufacture of such compounds. Lead compounds used in gasoline, although organic, are not pollution hazards in the organic form as they are changed to inorganic forms during gasoline combustion. They get released into the atmosphere as these less hazardous compounds. Inorganic lead gets absorbed primarily through the gastrointestinal and respiratory tracts and is the main source of lead for the body.

The major sources of lead poisoning today have been still old paint and industry. Solutions to these problems include education programs directed at the hazards of old paint, strict housing codes to prevent occupancy of hazardous dwellings, and better enforcement of existing safety standards directed at industry.

Lead is a cumulative poison. About 90% of the lead retained in the body enters the bones, from which it can be remobilised. The WHO recommends a maximum intake of 3 mg lead per person per week. Children and infants, who aae more susceptible than adults to lead poisoning, should have intakes of less than 1 mg Pb per week. The average rate of absorption of dietary lead is about 8%, but about 40% of the fine particulate lead retained in the lungs is absorbed (Fig. 3.5). These absorption rates result in about two-thirds of absorbed lead coming from the diet and one-third from the atmosphere. In addition

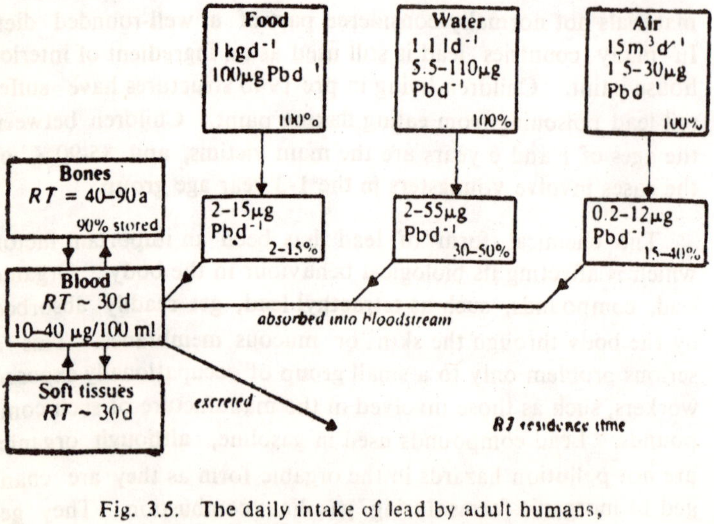

Fig. 3.5. The daily intake of lead by adult humans, and its distribution in the body.

lead intake is increased by about 5% for every 20 cigarettes smoked per day. The absorbed lead enters the bloodstream, where over 90% is bound to the red blood-lead cells with a mean residence time of one month. The blood-lead level appears to vary linearly with exposure and is therefore a good indicator of recent exposure, with normal levels being 10-20 μg Pb 100 cm^3 blood. 25-40% of the lead in blood enters soft tissues, about 15% enters bones, and the remainder is excreted. The residence time in the soft tissues is again about one month, with the majority of the lead being lost in hair, perspiration and digestive secretions. The concentration in hair is a good indicator of exposure to lead over the previous few months Bones act as the major reservoir for lead in the body and the residence time is 40-90 years in adults. Children absorb a great proportion of ingested lead than do adults, and the distribution in the body is also different. About 28% of the lead in children is in soft tissues compared with only 5% of the lead in adults being in the soft tissues.

Lead binds strongly to a large number of molecules, such as amino acids, haemoglobin, many enzymes, RNA and DNA; it thus disrupts many metabolic pathways. The effects of lead toxicity are very wide ranging and include impaired blood synthesis, hypertension, hyperactivity and brain damage. Effects at the molecular level, such as the interference with steps in haem synthesis, occur at very low levels equivalent to 20-200 μg Pb kg^{-1} soft tissue, which are the levels normally found in adults. If the normal body burden in adults is of the same order as that at which harmful effects may occur, any factor that increases lead intake must be viewed with great concern. The widespread distribution of lead from motor-vehicle exhausts increases atmospheric levels by factors of 20 (much more in urban areas); further, the subsequent contamination of soil and crops increases the amount of lead in food. Children, as well as absorbing a higher proportion of ingested lead than adults, have poorer hygiene habits and consume more dirt and dust picked up on their hands. In urban areas roadside dirt can contain several thousand parts per million lead (typical mean values 1000-3000 μg kg^{-1}) and samples of old paint in West London contained from 6.2% to 52.4% lead. Some children eat dirt, a habit known as pica, and they are particularly at risk from lead paints, one piece of which may contain enough lead to cause acute poisoning. Acute poisoning produces a rapid onset of symptoms due to the uptake of a large dose of toxic material. Chronic poisoning is due to the uptake of a large dose of toxic material. Chronic poisoning is due to the uptake of smaller doeses over a longer period of time and generally brings about a gradual appearance of symptoms. Associated with chronic poisoning are sub clinical effects in which there are no obvious outer (c: clinical) sings of adverse effects, but metabolic functions at the molecular level are being disrupted to a greater or lesser extent. Because of the lack of obvious symptoms, sub-clinical effects are difficult to diagnose. There is a growing body of evidence that the intelligence of children is being lowered in urban areas in which there are high inputs of lead from motor vehicles.

Blood lead levels are commonly used to monitor the status of lead body burdens. The results are sometimes not very satisfactory because there is not always a close relationship between blood-lead levels, soft tissue levels or concentrations that cause metabolic dis function. One individual with 25 μg Pb per 100 cm^3 blood may be suffering from lead-toxicity effects, whereas another with 70 μg Pb per 100 cm^3 blood may not. As it is individuals, rather than the statistical average members of the populations, that must be treated, a monitoring programme should ideally pick out the individuals at risk – otherwise either individual suffers will be missed or resources will be wasted on the treatment of too many unaffected people. However, the use of blood sampling will continue as there is no other sampling technique that yields better results and is so easy to carry out.

Arsenic

Methyl arsenic compounds occur in several water bodies in substantial percentages. But these form a small percentage of total. As in sea water. Some typical environmental data for As are given in Table 3.8.

TABLE 3.8

Environmental Analysis for Arsenic (μg/L)

1. Natural Water Bodies

Location	*As (III)*	*As (V)*	*Methyl arsenic acid*	*Limethyl arsenic acid*	*Total*
Fresh Water					
River	<0.04	0.18	0.08	0.36	0 66
Lake	0.90	0.50	0.24	0.19	1.83
Pound	<0.04	0.36	0.16	0.64	1.20
Sanine Water					
Tampa Bay	0.15	0.46	<0.04	0.26	1.91
Seawater, Bombay Beach					
Surface	0.019	1.59	0.08	0.24	1.857
25 m	0.018	1.34	0.005	0.16	1.523
100 m	0.08	1.69	0.004	0.006	1.78
Rain					
Dehradun	<0.003	0.20	<0.004	0.036	0.244

Arsenic speciation in air samples is carried out using a glass-wool filter to trap the particulates, followed by silvered glass beads to trap volatile arsines. The arsines are removed by dil. NaOH wash and analyzed as above, after $NaBH_4$ reduction. Ambient air is said to having As mostly as the inorganic species, some $(CH_3)_3$ As in vapour form, and some methyl arsenic compounds as particulates.

The speciation of As in human urine of Indians has been in Table 3.9.

TABLE 3.9

Arsenic Speciation in Human Urine of Indians

Species	*Average values*	*μg/L*
As(III)	1.5	2.1
As(V)	1.5	4.1
Methylarsonic acid	4.6	2.0
Dimethyl arsenic acid	12.5	16.0

The environmental chemistry of As is given in Fig. 3.6.

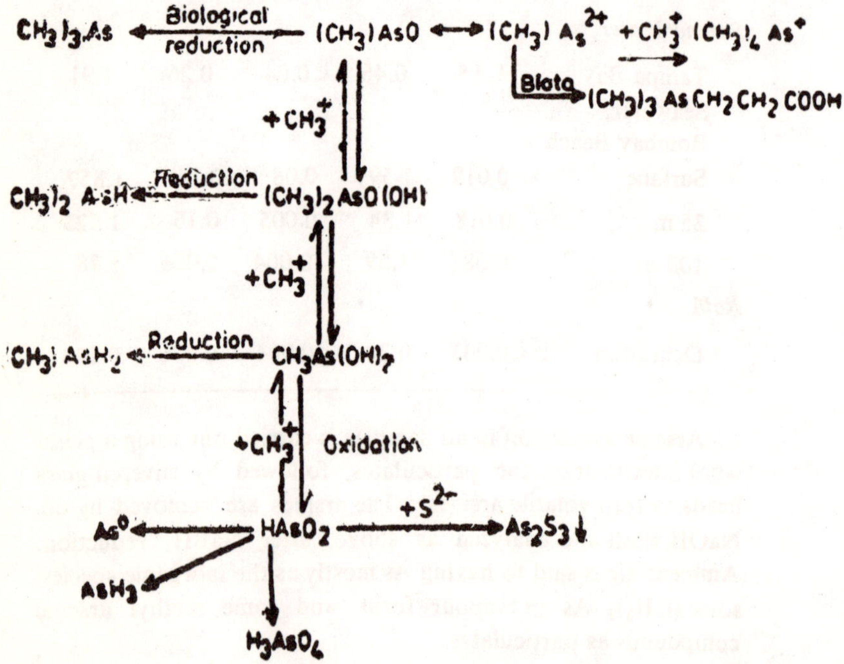

Fig. 3.6. Environmental chemistry of arsenic (R.S. Braman, "Arsenic in the environment in arsenical pesticides" ACS Symposium Series No. 7, American Chemical Society, p. 108-123 (1975).

Typical results of environmental Se analyses are depicted in Table 3.7.

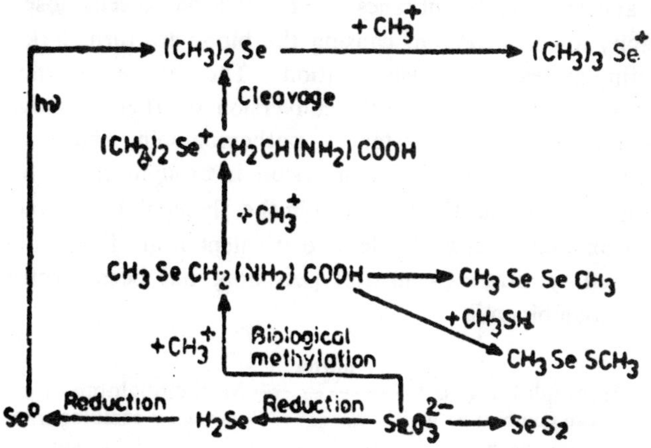

Fig 3.7. Environmental chemistry of Selenium.

Nitrates

The danger of drinking water containing high nitrate content has long been recognized. Livestock have been affected by well waters containing 75 to 150 ppm (parts per million of nitrate nitrogen. Human infants have been especially susceptible, not only because of their high gastic pH (low acidity) but also due to their high fluid intake relative to body weight. Nitrate poisoning brings about a condition called methemoglobinemia, resulting from changes in the hemoglobin of the red blood cells that reduce their capacity to carry oxygen.

Nitrate poisoning of infants has been reported from drinking water having nitrate in the range of 15 to 250 ppm of nitrate nitrogen (N), equal to 67 to 1,100 ppm nitrate ion (NO_3^-).

The primary source of excessive nitrates in well water has been the leaching of nitrate salts into the underground water supply from agricultural fertilization and the see page of sewage.

One of the symptoms of nitrate poisoning has been described as cyanosis, in which there has been a bluish discoloration of the skin and mucous membranes. The red blood cells lose their capacity to carry oxygen, causing the blood to turn dark and resulting in death from asphyxiation. The failure of the red blood cells has been due to the conversion of their hemoglobin to a closely related substance, mathemoglobin which cannot combine with oxygen. The conversion is brought about by the action of nitrite on the iron (Fe) in the hemoglobin molecule, which oxidizes it from divalent to trivalent iron. The condition produced by nitrate (or nitrite) poisoning has been termed as methemoglobinemia.

$$\underset{\text{can combint with oxygen}}{\text{Hemoglobin }(Fe^{2+})} \xrightarrow{\text{nitrite}} \underset{\text{cannot combine with oxygen}}{\text{Methemoglobin }(Fe^{3+})}$$

For a nitrate to be highly toxic it must first get converted to nitrite. In cattle it has been believed to be brought about by action of certain micro-organisms prevalent in the rumen, the first pouch of the stomach.

$$\underset{\text{nitrate}}{NO_3^-} \xrightarrow[\text{reduction}]{\text{microbial}} \underset{\text{nitrite}}{NO_2^-}$$

Nitrate poisoning takes place in many animals, including cattle, sheep, horses, turkeys, and man. Human infants are highly susceptible to nitrate poisoning. It is believed to get caused by the lower gastric acidity which favours nitrate reduction. The makeup of the intestinal microflora might also be a factor as well as the high carbohydrate diet of infants which favours the microbial action.

Nitrates have been normally present in many plants, often at levels that cause poisoning of livestock. Plants having 1 per cent nitrates have been toxic to cattle and 0.5 per cent has been to be the maximum that can be tolerated. Excessive fertilization is often the cause of high nitrates in plants. Treatment with hormone type weedkillers such as 2, 4-D can also cause high nitrate levels.

The blood of heavy smokers is having as much as 7 to 10 per cent of the hemoglobin converted to carboxy-hemoglobin because of habitual inhalation of carbon monoxide. Carboxy-hemoglobin is also not capable of transporting oxygen. so that effects of nitrate poisoning and carbon monoxide poisoning have been at least additive. The high concentrations of carbon monoxide in the atmosphere of urban and industrilzed areas adds appreciably to the burden of the oxygen carrying function of the blood. There has been also some nitric acid in the air from urban pollution, and nitrate salts are present in aerosol form. The contribution to nitrate poisoning has been probably minor, but when added to the rest of the burden, may be significant.

The effect of nitrates on the human fetus has not peen fully studied. When dregnánt cattle graze on land where the plants has been high nitrites. they may abort without showing any of the acute symptoms of poising. Adortion can be expected in pregnant animals that survive prolonged and severe poisoning. When fetuses get aborted in the absence of actuate poisoning symptoms in the mother, lesions that would be expected from prolonged hypoxia (oxygen starvation) are found in the fetus. Microscopic examinations show thickening and blocking of the arterioles. More information is required on the sensitivity of fetal and infant hemoglobin to the action of methemoglobin-forming agents.

A relatively recent cause for concern has been the potential cellular toxic effect of nitrates and nitrites. Substances that adversely affect cells are said to by cytotoxic. Nitrous acid, which contains nitrite nitrogen, has been known to be mutagenic since 1953. Its action is believed to be through the formation of nitrosamines.

Oil

Oil pollution has been an almost inevitable consequence of the dependence of a rapidly growing population on oil-based technology. The use of natural resources such as oil on a grand scale, without losses, has been almost impossible. The extent of such losses, intentional or accidental, has been steadily increasing and has been becoming a great cause for concern.

It has been estimated that the total oil influx into the ocean has between 5-10 million tons annually. The major sources are :

1. Cargo tanker washings at sea. The widespread practice of using sea water as ballast for empty tankers contributes an estimated 3 million tons of oil to the oceans annually. This occurs when the ballast gets dumped and carries residual oil from the tanker into the sea.

2. Bilge pumping at sea. The dumping of bilge contents by ships other than tankers contributes an estimated 5,00,000 tons per year.

3. Import oil losses. These loss take place during loading and unloading procedures and also as a result of collisions in port. The estimated annual contribution to oil pollution of the sea has been 1 million tons.

4. Other sources. Some other sources are not easily assessed but do contribute to the problem. Vessel accidents on the high seas or near shore certainly add to the problem and receive great amounts of publicity. The largest spill to date in this category occurred when the tanker Torrey Canyon ran aground off the coast of England in March to 1967. The oil spill amounted to 100,000 tons.

Losses also take place during exploration for and production of oil. The blowout of wells, disposal of drilling muds and other wastes, and accidental damage to offshore drilling rigs all constitute significant potential pollution sources. In this case, the flow of oil continued unabated for eleven days despite efforts at containment.

Oil leakage from the 200,000 miles of pipeline that crosses waterways and reservoirs has been a possible source of polluting oil. These lines have been subject to cracks, punctures, corrosion, and other sources of damage which would lead to leaks.

It is estimated that nearly 2 million lons of used lubricating oil is unaccounted for each year. It has been assumed that much of this eventually reaches coastal waters.

The Fate of Spilled Oil

Oil and water don't mix : so we have been told. According to this old adage, it might be expected that spilled oil would just float around until it got washed ashore. This, however, is not the case. One gallon of oil will cover an area of four acres of water when completely spread out. During the time it has been spreading, many ceanges take place in the oil. All oils, regardless of type, have volatile components that evaporate readily. Within a few days 25% of the volume of a typical oil spill is lost through evaporation. The remaining oil has been subjected to emulsification processes which, in a sense, do cause oil and water to mix.

Two types of emulsions have been formed. Oil-in-water emulsions result when small droplets of oil get dispersed in water and are stabilized by chemical interaction with the water at their surfaces. This process occurs especially well in turbulent seas. The resulting oil droplets do not get dispersed on top of the water, but rather throughout the water. Some oil droplets, especially those weighted down by mineral particles, even reach the bottom.

Water-in oil emulsions form when water dropletsget enclosed in sheaths of oil. The emulsion gets stabilized by interactions between the enclosed water droplets and various resinous and asphaltic materials found in most crude oil. This type of emulsion shows up as a floating, sticky, viscous mass that, on occasion, contains enough water to cause the total volume to be greater than that of the original oil.

Most oil that survives the emulsification process gets degraded by spontaneous photo-oxidation and oxidation by micro-organisms have been the most powerful source of oil decomposition found in the sea.

By the end of three months at sea only about 15% of the original volume of an oil spill remains. This is in the form of black, tarry lumps of a dense asphaltic substance, and it has been these lumps that frequently wash up on beeches.

When massive spills take place close to shore, sufficient time has been not available for the process described above to affect the total amount of oil involved. Under such circumstances a thick sticky film of oil gets deposited on any solid that comes in contact with the spill.

Biological and Physical Effects

The problem of oil pollution may be thought of in terms of short-term and long-term effects. The short-term effects have been immediately obviovs and have been the ones that have received the most publicity. The long term effects have been only slowly becoming apparent and have been currently the subject of much study.

Short-term effects fall into two categories, those caused by coating and asphxiation, and those that result from oil's toxicity. The effect caused by coating and asphyxiation are :

1. Reduction of light transmission. Measurement have regarded that ambient light intensity 2 metres below an oil slick was 90% lower than at the same depth in clear water. Photosynthesis by marine plant life may get hampered by extended periods of such conditions.
2. Dissolved oxygen reduction. Oil films are able to retard the rate of oxygen uptake by water. Measurements have shown a significantly lower dissolved oxygen level in water beneath a slick than in clear water. The final effects of this could not be determined.
3. Damage to water birds. Swimming and dividing birds get covered with oil. This causes their feathers to mat together, reducing their buoyancy and preventing flight,

The insulative value of the feathers has been also lost, and some birds quickly die of exposure in cold water. It has been estimated that 25,000 sea birds died because of the Torrey Canyon incident.

4. Smothering. The effects of oil plant life along shorelines are becoming a matter of concern. Intestidal algae and lichens are known to have been killed by smothering coats of oil which washed ashore in some areas.

The toxic effects of oil get resulted from some of the many compounds that make up this very complex mixture. These compounds have been found to very widely their molecular weights and structures.

Until recently, low boiling saturated hydrocarbons were regarded to be harmless to a marine environment. It has now been demonstrated that these substances, in low, water concentrations, cause anaesthesia and narcosis in a wide variety of lower animals. Cell damage and death result in animals exposed to high concentrations. It has been indicated that larvae and other young forms of marine life may yet especially affected.

Low-boiling aromatic hydrocarbons have been abundant in oil and represent it most dangerous fraction. Such compounds as benzene, toluene, and zylene have occurred in this fraction. These substances have been aute poisons for man as well as for all other living things. Naphthalene and phenanthrene, also present in oil, have been even more toxic to fish than the three compounds mentioned.

The aromatic comounds have been more water-soluble than the saturated hydrocarbons. The aromatics are able to kill marine organisms by direct contact (with oil) or through contact with dilute solutions created when the compounds dissolve into the water from the oil. The effect of these compounds diminishes somewhat with time because they have been volatile components of oil and evaporate.

Observations conducted on oil spills close to shore reveal that massive, immediate destruction of marine life takes place during the first few days after a spill. The species affected include a wide range of fish, shellfish, worms, crabs, and other invertebrates. Bottom-living fish and lobsters get killed and washed ashore. The toxicity has been immediate and causes death within minutes or hours after contact.

The long-term and especially low-concentration effects of oil components on living systems have been not as apparent as short-term effects. Some possible areas of concern have been proposed and are being studied.

Many biological processes of importance to the survival of marine life and occupying key positions in their life processes have been mediated by extremely low concentrations of chemical messengers in sea water. Marine predators, for example, have been attracted to their prey by organic compounds present in the sea water at concentrations in the ppb range. Similar chemical attractions and repulsions play vital role in such processes as finding food, escaping from predators, locating habitats, and sexual attraction.

There is reason to believe and some compounds of polluting oil have been found to interfere with these processes by blocking the teste receptors of organisms or by mimicking the natural stimuli, which gives rise to false responses on the part of the organisms. High-boiling saturated and aromatic compounds have been the ones likely to interfere in this way. Such interferenced could be having disastrous effects on the survival of some marine species.

The fate of organic compounds which enter the marine food chain has been studied. Results indicate that hydrocarbons, once incorporated into a particular organism, have been quite stable, regardless of their structure. There has been also evidence that these compounds can pass through many members of the marine food chain without becoming altered. Such compounds have been thus subject to food-chain amplification analogous to to the situation found for heavy metals and pesticides.

The food chain eventually reaches marine organisms that have been harvested for human consumption. This presents the possibility that the flavour of sea-food might be negatively affected, or, much worse still, the potential long term positions might accumulate in food used for human consumption. This latter point has been especially unnerving when it has been regarded that some of the high-boiling aromatic hydrocarbons involved have been known or thought to be carcinogenic.

The food-chain amplification could also give rise to situations in which marine life high on the chain would accumulate quantities of the compounds sufficient to destroy them

Emulsified oil may reach the sea bottom, as mentioned before, especially if it is weighted down by mineral particles. Oil on the sea floor usually persists for long periods of time and ban continue to damage bottom plants and animals. The destruction of these plants and animals is able to reduce the cohesion of bottom sediments and accelerates their tr nsport. Sediment movement has been commonly observed following oil spills. Contaminated sediments can be spread over great distances under the influence of tide and wave action.

Recent reports have posed an additional environmental threat from oil pollution : the oil may serve as a concentration medium for other fat soluble poisons such as pesticides. Dissolved in an oil film, these poisons might reach concentrations many times higher than those common in polluted. These poisons could, in this way, reach susceptible organisms in concentrations much higher then they would normally encounter.

Countermeasures for Oil Pollution

The obvious solution to this problem has been to prevent it from happening. As the principal cause of oil spills has been human error, much effort can and should be directed at prevention. An area where progress can be made in this direction is transportation, where improved navigational aids (both ship- and shore- based) and improved safety and hazard warning instrumentation could prove effective. Off-shore drilling operations require improvement. Some suggestions have been the development of submerged pyramid shaped canopies to cover the drill hole area and collect and trap any spills, the use of mechanical or pneumatic (air curtain) walls around the drill site, and the use of some form of physical encapsulation of the drill and drill hole.

Accidents will occur in spite of prevention methods, and preparations must be made to handle such events. This is especially crucial because accidents often cause in massive amounts of oil being released into small areas. The areas of immediate concern in the event of a massive spill have been containment of as much oil as possible at the scene of the spill, and countermeasures for any oil that escapes containment. At the present time, neither area has been without numerous problems.

The containment of an oil spill of necessity has been involving the creation of some sort of barriers. Three types of barriers are used or are in the developmental stage. They have been floating booms, water bubble barriers, and chemical barriers.

Floating booms have been in common use in harbors and areas where the transfer of petroleum products occur. These booms have been constructed of a tubular floating section, which is inflatable or is filled with some buoyant material. Some sort of weighted skirt is generally suspended below the floating section. These booms are used to encircle and contain spills until they can be removed.

The success of these devices has been found to depend upon the conditions under which they get used. They are most effective in calm water and when used on small spills. Most commercially available models have been unsatisfactory in the open sea or in any rough water because they become difficult to anchor, and wave action carries the oil over and under the barrier. Large, more bulky and more rigid booms overcome this problem, but they have been more difficult to transport and set up.

In emergencies, make shift booms have been used. Construction materials for these devices include inflated fire hose, linked railroad ties, and telephone poles. A considerable amount of development and improvement has been necessary to make floating booms useful in the rough-water situations that often accompany (and sometimes cause) accidental spills at sea or in harbors.

Water bubble barriers could be formed by releasing compressed air from a perforated, submerged piece of tubing. The air rises to the surface as a curtain of tiny bubbles. The resulting agitation disallows or at least inhibits the movement of oil films across the curtain. The barrier works as long as water current and wind do not overcome the forces set up by the compressed air. These barriers have been quite new and although totally impractical for the open sea, they are in use in some harbors. One nistinct advantage has been that they do not impede the movement of vessels.

Certain chemical additives have been capable of gelling or even solidifying oil. If these could be applied effectively to the periphery of an oil slick, it has been conceivable that a ring of gelled oil could serve as a containment barrier, possibly in very rough seas. Work along this line has been still in the experimental and development phase.

The job of containment has been much easier in harbers and protected waters than on the open sea, where high waves and anchoring still does not allow effective use of barriers.

Whether or not containment of a spill has been successful, the task of cleanup and removal must get accomplished. The rash of spills in recent years has allowed numerous cleanup methods to be tried with the hope of finding a "best" one. Apparently some methods have been better than others, and, in fact, some should be avoided completely because of long-term effects. Indeed, some of the more obvious methods have been among the least desirable.

Mechanical methods of cleanup have been very satisfactory from two points of view. The oil has been removed from the water and reclaimed for productive use. Mechanical removal involves the use of skimmers, which get attached to ships and which effectively skim the oil and a thin layer of water from the surface. The oil and water are allowed to separate in storage. tanks within the ship, and the water is then pumped out.

All skimmers now in use work best in calm waters and on thick layers of oil. They are quite effective on oil slicks contained by floating booms. The present recovery rate of skimmers has been between 10 and 50 barrels per hour, depending on the calmness of the water and the type of skimmer. In summary, skimming has beeen an ideal method for use on small, contained spills, but it has been beset with difficulties when used on large spills in open water.

Hydrocarbons have been naturally degraded by marine microorganisms. Accelerated biological degradation, using these same microorganisms, could be applicable to those oils spills in the open sea which will not reach shore for several days. This technique would be involving bacterial seeding and fertilization of the oil slicks. In order to assure complete degradation, sufficient time would be required and all appropriate bacterial species would have to be used. These bacteria are highly selective, and no single species could degrade all components of the oil.

A number of problems have been found to accompany this technique. The main one has been the severe oxygen requirement for the degradation process. The complete degradation of 1 gallon of crude oil requires all the dissolved oxygen in 320,000 gallons of air-saturated sea water. Therefore, the oxidation will be slow (and may be the slick will reach shore) in areas where the oxygen content of the water has been low (possibly from previous pollution). The degradation might bring about additional damage to marine life by depleting the oxygen.

Combustion was one of the earliest methods used to rid water of oil pollution. It has been an attractive and inexpensive method for disposing of large amounts of oil. It has been. of course, inadvisable in situations where the oil has been close to shore or to structures that could be damaged.

Generally, burning has been effective only if it can begin immediately after a spill. The lighter fractions of oil evaporate immediately after a spill, and the remaining oil is having a higher ignition temperature and can be difficult to ignite. This factor, coupled with the ability of water to conduct heat away from the burning oil, has caused some attempts to be unsuccessful. Research has been now aimed at developing "aids to combustion" in the form of wicks, catalysts, and supplementary fuels. Further developments in burning techniques may give rise to methods very useful on large spills in the open sea. One drawback has been that water pollution has been converted to air pollution, since dense black smoke is generated. The residue after burning has been a black solid that has been easier to collect than the liquid oil.

Small amounts of oil could be removed by sorbents. Oil gets soaked up by an absorbent and clings to the surface of particles of an adsorbent. Either type of sorbent could be used to remove oil from the surface of water. The use of such materials is having the advantage of limiting the rate of slick spreading and facilitating cleanup. The disadvantages include problems of delivery and application of the materials, and collection and disposal of the oil-soaked products. Also, the method is prohibitive with extremely large slicks due to the amounts of materials involved.

Straw and sawdust are used as sorbents. After becoming saturated with oil, they might be removed and disposed of. Shredded polyurethane foam, an absorbent, is used and has the advantage that, when it gets saturated, it can be removed and the oil can be squeezed out. The foam can then be reused. The recovered liquid is 80% oil, which can be recovered for use. A volume of 50 ft^3 of foam will reportedly soak up 1 ton of crude oil.

Sinking agents, in the form of sand, brickdust, cement, silicone mixtures, and fly ash, are added to oil spills. The combination of oil and sinking agent has been heavier than water, and the oil sinks. Some of the agents have been chemically treated to increase their properties of being attracted to oil and rejected by water. In order to be most effective, the sinking agent should not release the oil back to the water environment. This has been a problem with some agents.

The French used large amounts of sinking agents on the soil slick from the Torrey Canyon incident. It removed the immediate problem of having the slick reach the shorelines.

It now appears, because of studies on long-term effects, such as those previously discussed that physical sinking has been one of the least desirable methods of treating oil spills. Sunken oil kills bottom-dwelling animals so rapidly that most mobile dwellers have insufficient time to move. Oysters, scallops, and lobsters have been among the species affected. The oil may form a layer or "blanket" on the bottom and persist for long periods of time.

The bacterial degradation of sunken oil needs large amounts of oxygen. Because of this sediments loaded with oil may become anaerobic. One of the principles of geochemistry has been that hydrocarbons in anaerobic sediments persist indefinitely. Anaerobic sediments have been very slowly reworked and repopulated, so such sites might constitute sources of water pollution for a long time after the original spill.

Many materials have been available that will be able to emulsify oil in water. These products are known by a variety of names, such as emulsifiers, detergents degreasers, and dispersants. We shall employ the term dispersant. The primary components of a dispersant have been a surfactant, a solvent, and stabilizers. The surfactant has an affinity for both oil and water and allows the two to mix. Solvents enable the surfactant to penetrate into the oil slick. The solvent usually makes up the bulk of the dispersant composition. Stabilizers are able to fix the emulsion once it is formed. The net effect of the dispersant has been to lower the surface tension of the oil to the point where it will break up and disperse in the water in the form of tiny droplets.

The action of dispersants is able to increase the surface area of an oil slick and distributes the oil droplets into a large volume of water. These two effects are able to enhance the natural degradation by microorganisms. The chemical dispersants themselves are not able to destroy any oil.

Dispersants have been usually applied by spraying. Agitation becomes necessary to make them effective. This makes them very useful on oil slicks in rough open sea where nature does the agitating. Evaluation of the results of their use in the past reveals that dispersants must be used with caution. The high degree of dispersion they create leads to greater solubility and higher concentrations of oil in the water than would be possible if natural dispersion were allowed to take place. Also, the oil droplets formed have been in a size range that has been easily ingested and assimilated by marine organisms on the low end of the marine food chain. In the case of the Torrey Canyon incident, some of the dispersants used were found to be as toxic to aquatic life as the oil itself.

Dispersants used on beaches actually add to the problem by making the oil to penetrate more deeply into the sand. The compactness of the sand has been upset, and it becomes more easily moved by tide and wave action. This contributes to beach erosion.

Acid Mine Drainage

It is a very common and dangerous problem in aquatic environment which is somewhat similar to acid rain problem.

Coal mines discharge appreciable quantities of H_2SO_4 and also $Fe(OH)_3$ into streams through see page. These are formed by the oxidation of FeS_2 in pyrite which is found in large quantities in the underground seams which are having coal. FeS_2 is known to be stable in absence of air but when the coal seams get exposed to air in mining operations, the oxidation reactions occur as shown below giving large quantities of acid. These reactions tend to continue long after the coal mining operations are over. Micro-organisms play important role in the overall process which involves several reactions.

$$2FeS_2 + 2H_2O + 7O_2 \rightarrow 2Fe^{2+} + 4HSO_4^- \quad ..(1)$$

$$4Fe^{2+} + O_2 + 4H^+ \rightarrow 4Fe^{3+} + 2H_2O \quad ..(2)$$

The above reaction (2) has been slow at pH below 3.5 but has been catalyzed by the iron bacterium, Thiobacillus ferroxidans. At pH 3.5 - 4.5 the reaction could be able to catalyze by a variety of bacterium include, Metallogenium. The other bacteria involved in the process include thiobacillus thiooxidans and Ferrobacillus ferrooxidans. Fe^{3+} dissolves pyrite further and together with the second reaction (2) above, forms a cycle for the dissolution of pyrite.

$$FeS_2 + 14Fe^{3+} + 8H_2O \rightarrow 15Fe^{2+} + 2SO_4^{2-} + 16H^+ \quad ..(3)$$

$Fe(H_2O)_6^{3+}$ has been acidic which at pH above 3 yields a precipitate of $Fe(OH)_3$:

$$Fe^{3+} + 3H_2O \rightleftharpoons Fe(OH)_3 \downarrow + 3H^+$$

The stream beds which are contaminated with acid mine water, have been generally coated with odd yellow deposit of amorphous semigelatinous $Fe(OH)_3$. H_2SO_4 of acid mine water has been able to destroy aquatic life in water bodies

The prevention of water pollution from acid mine water is not as easy job for environmental chemists. Carbonate rocks could be used for remedial measures :

$$CaCO_3 + 2H^+ + SO_4^{2-} \rightarrow Ca^{2+} + SO_4^{2-} + H_2O + CO_2 \quad ..(4)$$

But with the increase in pH as this reaction proceeds, $Fe(OH)_3$ occurring in the system tend to cover the particles of carbonate rock with a relatively impermeable layer. This is able to inhibit further neutralization of H_2SO_4.

3.7 Pesticides. See separate chapter for this.

3.8 Soaps and Detergents. See separate chapter for this.

4

Analysis of Water Quality

Introduction

Water pollution study includes the estimation of physical, chemical and biological parameters for which a medium size, well equipped laboratory is needed. The number of parameters to be determined depends, by and large, upon the purpose to be achieved. For example, if water under examination is intended to be used for public water supply and bathing, bacteriological investigation is a must. If the water is planned to be used for fish culturing, organic carbon, dissolved oxygen (DO), and NPK (nitrogen, phosphorus and potassium) ratios are to be estimated on priority basis. If the water is to be used for boiler and in food manufacturing industries, then, hardness, alkalinity, iron, manganese and silka must be thoroughly investigated, otherwise many problems are likely to obstruct the production. Keeping in view the requirement of water in promoting industrial progress, maintaining public health, boosting up of agricultural production and almost in every walk of life methods for estimating altogether 30 different physico-chemical and biological parameters have hereinafter been briefly discussed so that the suitability of water for a variety of purposes be easily adjudged. For correct estimation of the parameters, proper methods of sampling and purity of the chemicals used in the analysis are equally important Special care is needed in water sampling from different parameters are discussed below.

Sampling

The accurate assessment of the concentration of pollutants present in water primarily depends upon the samples drawn for the analysis. The samples must be representative in nature. The samples are generally of three type :

1. Grab samples are those which are drawn from a fixed point. Such samples reflect the existing physico-chemical condition in the system with respect to the time of the withdrawal of the samples.

2. Composite samples are those which are collected from various parts of the same system for examination, over a fairly extended period. Such samples reveal the overall characteristics but do not reflect the anoxic conditions, existing in aquatic ecosystem, at the time of the sampling.

3. Diurnal samples are those which are drawn in the grab manner from the fixed sampling sites at regular time intervals in a single day. Thus, the changing condition of the existing aquatic ecosystem, with respect to the varying intensity of solar radiation is known by this method. This type of sampling is more desirable in limnological studies where production is to be evaluated.

Frequencies and Number of Samples

The frequencies, duration and number of samples drawn, depend upon the purpose and the source from where the samplings are to be done. Supposing if diurnal variation in a pond or stream is intended to be measured, six hourly samples need to be drawn at least for complete 24 hours. Similarly, if potable water samples are to be drawn for bacteriological analysis from the public water supply establishment in a town of more than one lakh population, samples are to be drawn daily. If the number of persons ranging between 50,000 to 100,000 then sampling is required twice a week, whereas for population ranges between 20,000 to 50,000, samples of water need to be drawn once in a month. In case of small turbulent stream two samples are sufficient, one from upstream and another from downstream In cash of shallow lakes or ponds, water samples from 30 cm and 0.5 m depth are required to be drawn for bacteriological and chemical analysis respectively. In cash of deep lentic water bodies, several samples are drawn vertically. The number of water samples to be drawn depends by and large on the area of the pond or lake to be studied. If these are 0.4 km^2 in area, only one sample is required, but if the area covered is more than 1.0 km^2 then a minimum of 3 wate samples must be drawn. The quantity of samples to be drawn for plankton analysis depends on the approximate number of planktons present in the system. If the ponds are oligotrophic, 25 litres of water should be sieved through the suitable size of the plankton net. If the ponds happen to be eutrophic, a minimum of 5 litres of water samples are to be sieved through the net.

Site of Water Sampling

So far as the site of water sampling is concerned, too near ro too distant sampling points should be avoided for chemical analysis. Stagnant zone of the surface water should be avoided for bacteriological samples. The water samples should always be collected against the current and in cash of persisting stagnation, an artificial current should be created while samples are being drawn for bacteriological studies. On the contrary, the water samples should always be drawn from the clam-zone of the rivers and ponds for identification and evaluation of the plankton.

Time Interval Between Collection of Water Samples and their Analysis

Some of the parameters such as the determination of pH, temperature, dissolved oxygen, free carbon dioxide and many other dissolved or evolving gaseous constituents must be analysed on the spot as soon as the samples are collected. For the rest of the chemical parameters, samples must be brought to the laboratory in well preserved condition, as early as possible, but in no case the time interval between the collection of the samples and the beginning of the analysis, should exceed more than 72 hours. The samples collected for bacteriological studies must be subjected to analysis within 24 hours of their collection. The samples for the identification and evaluation of plankton must be preserved by adding 4 per cent formalin as preservative soon after the collection is over, perferably on the spot. Even then analysis is required to start at the earliest possible opportunity preferably within a month of collection of samples.

Methods of Water Sampling

The method of water sampling mainly depends on the nature of analysis to be carried out. Analysis may be broadly divided into physico-chemical, bacteriological and biological categories. The biological analysis includes the identification and quantitative estimation of phsio and zoo-plankton, benthos and other biota generally found in the aquatic ecosystem.

(i) Water samples for chemical analysis : The water samples are collected from a depth of 0.5 m in case from open water bodies and in case of piped water supply system, from a tap, in thoroughly cleaned jar or nutral glass containers of minimum 2.5 litres of capacity, provided with double cap device. The samples are collected upto the top, without leaving any space so as to prevent the premature release of dissolved gases during the transit period.

(ii) Water samples for bacteriological analysis : The water samples are collected from 30 cm depth in cash of open water bodies and for a public hydrant after sterilising the tap for a few minutes with the help of a spirit lamp, prior to taking the samples from the same in properly sterilized neutral glass bottles of 120 ml. capacity. To each of these sampling bottles, 0.1 ml of 30 per cent sodium thiosulphate solution is added prior to subjecting it to the process of sterilization. The sample of water should not be collected upto the top. On the contrary, some space must be left for bacteria to survive The collected water samples are brought to the laboratory from the field in the packed sampling box and the analysis should be taken up positively within 24 hours of collection.

(iii) Water samples for plankton analysis : 50 litres of water from the oligotrophic water bodies are sieved through plankton net, preferably of 0.1 mm mesh size. In case of eutrophic water bodies (plankton court more than 25,000/ml), no sieving is required. The algae so collected are immediately fixed by adding 1 to 2 ml of 4 per cent formalin solution as preservative. Alternative preservative has also been suggested for which the use of 1 ml of lugol's solution (prepared by dissolving 10 Kl, 5 g sublimed iodine and 5 g sodium acetate in 100 ml distilled water) is indicated.

(iv) Water samples for benthic biota : The biota living in the bottom mud layer of the lakes, ponds or rivers, reflect the overall qualities of the water column standing over the stratum. The most conveniently used instrument of collecting benthos in the limnological studies is Ekmann dredge by which the mud samples from unit area are drawn. The samples of mud so collected are applied along with water to the combination of sieves, put together one over the other. The sieve of mesh size of 0.5 to 0.6 mm is kept abve and another sieve of 0.2 m mesh size is kept below it. The biota separated are picked up with the help of forceps and put into a glass container of 50 ml capacity with wide mouth, to which 10 per cent formalin solution is added as preservative. The samples of the biota so collected are further subjected to analysis.

PARAMETERS ANALYSED ON THE SPOT

Hydrogen-ion-concentration (pH)

The hydrogen ion concentration (pH) is as important as the total solid content, for the sewage and temperature for ecological studies as a whole. Waste water with an adverse concentration of hydrogen ion is difficult to treat by biological means and if the concentration is not altered before discharge, the waste water can alter the pH of natural waters. In domestic waste water (containing sewage and septic effluents), the pH value normally varies from 0 to 7.5 and therefore do not affect the receiving water body but an altogether different picture may arise in the case where acidic or alkaline wastes go into the streams. Acidic wastes are corrosive to both metallic and concrete structures. Besides beirg toxic to the aquatic life, they react with the natural alkalinity of the water, thereby increasing the carbonate hardness and thus rendering it unfit for future uses in laundry or boiler. On the other hand, water having pH value of more than 8 when discharged into the receiving stream, combines with free carbon dioxide and further increases the alkalinity of the water. The optimum ranges of pH for aquatic life are 6.8 to 9.0.

Estimation

The pH value of clear water in the field is determined with the help of Lovibond pH comparator disc. Water samples are filled-up in the pH cell upto the marked calibration followed in quick succession with the addition of 3 to 5 drops of phenol red solution as indicator. After shaking the sample, the colour developed is matched in the disc by which the pH value is known. In case of coloured water sample or any coloured solution pH value is estimated by using electronic pH meter.

Dissolved Oxygen (DO)

The dissolved oxygen (DO) content shows the health and ability of the stream to purify itself through biochemical process. Too high content of dissolved oxygen is an indication of organic pollution, particularly when pollution is contributed by sewage outfall. When DO level drops below approximately 5 ppm, the more desirable species of fishes such as trout and brass leave the area and only the coarse type of fishes predodominate. Further it is reported that at an oxygen level of approximately 2 ppm, fishes disappear and the environment shifts towards anaerobic species. High depletion in oxygen content produces foul odour due to the anaerobic decomposition of organic waste leading to the formation of hydrogen sulphide.

Estimation

DO is estimated on the spot by adding 2 ml of manganese sulphate solution (prepared by dissolving 364 g $MnSO_4$, 4 H_2O, in 1 litre distilled water) to the freshly drawn water samples in glass bottles of 300 ml capacity. The content is mixed by swirline 2 ml of alkaline sodium iodide azide (prepared by dissolving 700 g of sodium hydroxide, 135 g of sodium iodide and 10 g of sodium azide in 1 litre of distilled water), solution is then added. The whole content is thoroughly mixed and the resultant brown flock is allowed to settle for about 5 minutes. Thereafter, 2 ml of conc. H_2SO_4 is added whereby the settled flock gets dissolved and makes yellow solution on mixing 200 ml of the above solution is taken out in another cleaned neutral glass bottle and titrated against N/40 sodium thiosulphate solution using starch as indicator. The amount of titrant used gives the direct reading for DO in ppm.

Temperature (°C)

Temperature is one of the most important parameters for aquatic environment because almost all the physical, chemical and biochemical properties are governed by it. For example, density, viscosity, surface tension and vapour pressure of water, more or less depend on the temperature profile of the system. It is the temperature which limits the saturation values of solids and gases that are dissolved in it. The rate of chemical reactions and other biological activity such as corrosion or incrustation, BOD, photosynthesis, growth and death of micro-organisms are all dependent upon environmental temperature values. Indian climate provides almost an ideal range of solar temperature which attributes a great self purification strength in the stream. The rate of biochemical reaction is directly proportional to the environmental temperature in hottest summer months, the oxygen demand increases, leading to serious oxygen depletion problem in the systems concerned.

Estimation

The temperature of water measured by using a thermometer, calibrated in °C scale. For measuring the temperature of open water body, the thermometer is dipped into the water and the reading is taken. Thermometer should not be taken out from the water body for reading the temperature. If the thermometer of °C scale is not available, then one of °F may be used.

Turbidity/Transparency

Turbidity of water is responsible for the light to be scattered or absorbed rather than straight transmission through the sample. It is the size, shape and refractive index of the suspended particulates rather than the total concentration of the latter present in the water samples that are responsible for turbidity.

Turbidity decreases light penetration, limit production of phytoplankton which consequently leads to decrease in photosynthetic activity and depletion of oxygen content. Clear ponds with less than 25 ppm turbidity may have 12.8 times more plankton and 5.5 times more fish production than ponds with a turbidity exceeding 100 ppm. High turbidity closes the door and impairs the lungs by completely blanketting the upper surface of water body. Since the surface not only acts as window through which it receives radiant energy but it is also like lungs through which it takes in oxygen from the atmosphere and gives out carbon dioxide and other dissolved gases that are generated within the water by microbes. The turbidity and the transparency values are interrelated, just like that of total dissolved solid and conductivity values. The transparency values are expressed in cm or mm and turbidity in ppm. The values of transparency and their corresponding values for turbidity (in ppm) are shown in the table on the next page.

TABLE 4.1

Relation between Transparency (mm) and Turbidity (ppm)

Transparency (mm)	*Turbidity (ppm)*	*Transparency (mm)*	*Turbidity (ppm)*
20	1000	32	600
22	900	34	450
24	800	100	100
26	700	328	25
28	600	390	20
30	550	490	10

Odour

Odour in water is caused by volatile substances associated with organic matter, living organisms principally algae and related organisms and gases. Offensive odour causes poor appetite for good food, lowers down the water consumption, impairs general quality and causes nausea and vomitting. Odour can result in marked decline in tax revenues, pay rolls and sales. World Health Organisation (WHO 1961) has recommended that water intended for drinking, bathing and recreational purposes should be completely free from any odour.

Colour

The dissolved or colloidal sabstances of vegetable origin which generally consists of tannins, glucoside as well as iron and other substances are responsible for the colour piesent in water. The colour of sewage reflects its strength and condition. Fresh sewage is grey whereas septic sewage is black in colour. The colour of river water is reduced by storage or ageing of the water and by bleaching action of sunlight.

Estimation

1 g of cobalt chloride along with 1.245 g of potassium chloroplatinate are dissolved in distilled water and the solution is then diluted to 1 litre when the same has a colour of 500 hazen unit. Maximum permissible colour for drinking purposes is 20 ppm on platinum cobalt scale. Colour is harmless but it cannot be allowed more than the maximum tolerance limit fixed as the consumers feel repellant.

PHYSIO-CHEMICAL PARAMETERS

Total Dissolved Solids (TDS)

TDS contents of water and waste water are defined as residue left upon evaporation at 103°C to 105°C. It is an aggregated amount of the entire floating, suspended, settable and dissolved solids present in the water sample. Further, in all the three above mentioned physical stages, the solids may be organic or inorganic in nature, depending upon volatility of the substances. The organic fraction is driven off as gas on ignition at 600°C and the inorganic fraction remains as ash. Thus, the term volatile and non-volatile solids refer respectively to organic and inorganic contents of the water sample. Organic constituents of total solids are responsible for taste, colour, odour, gas and biological problems, whereas most of inorganic materials being non-biodegradable do not cause such environmental pollution problems, although a few of them, in traces, produce acute toxicological problems for aquatic dwellers. Inorganic mineral contents, present in the water, conduct electric current which is measured by specific conductivity tests. An approximate idea of total dissolved solids (TDS) in natural water can be obtained by multiplying the specific conductance at 25°C by a factor of 0.55 to 0.75 with an average of 0.65. TDS in ppm can also be found out by adding the sum of the determined constituents in ppm except that the bicarbonate concentration should be multiplied by a factor of 0.49. Osmotic pressure in atoms is known on multiplying the conductivity value of mhos/cm by

0.00036. Studies carried out on the productivity of fishes with respect to conductivity values suggest that electric conductivity above 400 mhos/cm do not limit production. On the other hand, the fish production does not increase proportionately with conductivity.

Alkalinity

It is an anionic phenomenon. All anions such as CO_3^{-2} (carbonate), HCO_3^- (bi-carbonate), OH (hydroxyl), PO_4^{-3} (phosphate), SiO_4^- (Silicate) etc., contribute alkalinity to water. The number of mili-equivalents of acid used in the titration to combine all the hydroxyl ions, is known as total alkalinity. The water supplied with less than 103 ppm of alkalinity as $CaCO_3$ is ideal for domestic use. The ratio of alkalinity to that of alkaline earth metals is a good parameter for determining the suitability of irrigation water.

Estimation

To 100 ml of water sample, 2 drops of 0.1 per cent phenolphthaloin (P) solution, as indicator, is added and titrated against N/50 H_2SO_4 solution, until the pink colour disappears. Thereafter, another 2 drops of 0.1 per cent methyl orange solution as indicator is added to the same sample of water. Titration against N/50 H_2SO_4 is continued until the yellow colour changes to pink. The titrant consumed in ml give the value for total alkalinity (TA). The calculation for different types of alkalinity involves some mathematical interplay which is as such :

(i) When P = O, TA is HCO_3 alkalinity as $CaCO_3$.

(ii) When P = TA, TA is OH alkalinity as $CaCO_3$.

(iii) When P = 1/2 TA, TA is CO_3 alkalinity as $CaCO_3$ and equal to 2 P. But when P is either less or more than half of total alkalinity (TA), then altogether a different situation arises. More than one type of alkalinity exists in the same sample.

(a) When P < 1/2 TA, then CO_3 will be 2 P and HCO_3 will be T-2P.

(b) When $P > 1/2$ TA, then 2 P·T will be for OH and 2(1/2-P) will be for carbonate type of alkalinity.

Hardness

Hardness is caused due to bi-valent cations such as Ca^{++}, Sr^{++} etc. Higher cations also contribute hardness to a lesser degree but mono-valent cations never produce hardness. Thus, sodium compounds do not cause hardness from the agricultural, limological and industrial point of view. The estimation of hardness is of vital significance but has a very limited scope in respect of environmental considerations. Water should have zero hardness specially for boilers operating at high pressure. Limnologicaly, moderately hard water is desirable.

Estimation

To 100 ml of water sample, 2 ml of buffer solution (prepared by dissolving 16.9 g NH_4 Cl in 143 ml of concentrated ammonium hydroxide solution) is added to bring the pH value to 9.0. Then one small spoon (say 0.20g) of Erichrome black-T is added as indicator which forms wine red colour. The whole content is stirred and titrated against 0.01 M solution of EDTA (ethylene diamine tetra acetic acid), till the wine red colour changes to blue. To avoid the possibility of interference by trace elements, if suspected to be present in the water, one ml of 5 per cent sodium sulphide (Na_2S) solution is added as inhibitor, prior to the procedure for titration. The titrant consumed in ml on multiplying by factor 10 gives the value for the total hardness in ppm as $CaCO_3$.

Relation between Hardness and Alkalinity

When the total alkalinity valve is higher than the value for total hardness, then the latter indicates the carbonate type of hardness. If and when the total hardness is higher than the total alkalinity, then the value for total alkalinity, is for carbonate hardness and the difference of the two shows the value for non carbonate hardness.

Calcium

Calcium is one of the alkaline earth metals, the others being magnesium, barium, strontium etc. It is widely distributed on the earth's crust and is present nearly in all waters. In presence of CO_2, Calcium bicarbonate can normally be dissolved upto 20 ppm (as Ca) at atmospheric pressure and upto 100 ppm at higher pressure. The water coming from limestone zone may contain greater concentration of calcium. As soon as CO_2 is released, calcium carbonate (which can be retained in water upto 5 ppm as calcium only) is precipitated out. Lakes have been classified as poor, medium and rich, depending upon their calcium content, being less than 10 ppm, between 10 to 25 ppm and more than 25 ppm respectively. Calcium is not known to indicate or produce any hazardous effect on human health. Concentration upto 1800 ppm has been reported as not to impart any physiological reaction on human being. Moderate content of calcium in water is rather desirable because the toxicity of various substances *i.e.* Pb, Cd, Hg etc. have been observed to be neutralised. Undoubtedly, high concentration of calcium content in water is not desirable at all for laundering and for use in boiler. Calcium, if present with sulphate, inhibits malt fermentation and with chloride, inhibits growth of yeast in distillery industry.

Estimation

2 ml of 1N NaOH solution is added to 100 ml of water sample, to raise the pH above 10. At such high pH, magnesium is precipitated out as $Mg(OH)_2$ but Ca remains in solution. It is than treated against 0.01 M solution of EDTA, by using muroxide as indicator. The titration is continued till the solution becomes distinctly pink in colour. The titrant consumed in ml, on multiplication by factor 4, gives the concentration of calcium as Ca (in ppm). Calcium as $CaCO_3$ is obtained in ppm by multiplying the titrant value in ml by 10.

Magnesium

Its conceniration is known by direct mathematical calculation (in ppm) as Mg. The formula used is as such :

Magnesium as Mg in ppm = 0.24 × (Total hardness as $CaCO_3$-Calcium as Ca)

Magnesium has ten times the solubility of calcium and being bivalent, it too produces hardness. It is absolutely essential for the chlorophyll bearing bacteria, algae and plants. It is second to calcium in hard water, whereas sodium (Na) becomes second to calcium in soft water. Na is metabolized only by blue green algae but potassium (K) along with magnesium is a necessary requirement for all algae. Under low Mg and K conditions, growth and photosynthesis of algae are poor and respiration is high.

Chloride

The sudden increase in the concentration of chloride in surface water bodies at abnormally high concentration was previously used as index of pollution through contamination by fecal matters. Human excreta contains chloride equal to the chloride consumed with food and water. Normal human body discharges from 8 to 15 g (av 9 g of chloride) a day. Since chloride ions are non-biodegradable, these are not removed from the wastes even after subjecting it to the secondary treatment process. However, it is harmless upto 1500 ppm concentration but produces a salty test at 250-500 ppm level.

Estimation

100 ml of water sample is taken in white procelain basin to which 1 ml of 5% potassium chromate solution, as indicator, is added and titrated against standard silver nitrate solution (prepared by dissolving 4.79 g of $AgNO_3$ in one litre). The titrant consumed in ml on multiplying vy 10, gives the direct reading for chloride content of water in ppm.

Flouride

Flouride concentration in water both at levels less than 1.0 ppm and more than 1.5 ppm, are harmful for health. According to UNESCO (1963) and California State Water Pollution Control Board (1952), water containing more than 1.0 to 1.5 ppm of flouride can cause mottled enamel in children and an excess of flouride may eventually cause endemic cumulative flourosis with resultant skeletal damage in both children and adults. If the flouride concentration in drinking water is less than 0.5 ppm, a high incidence of dental carries is likely to occur. Recent researches by Geological Survey of India have led to the discovery of a natural mineral setpentine which can absorb large quantities of flourides from watar and can act as a deflouriding agent with considerable success in the treatment of flourosis.

Estimation

50 ml of water sample along with control solutions are taken in different nesselar tubes and put on the procelain based wooden stand. To each of the tubes, 2 ml of zirconium alizarine solution is added. 90 minutes after the addition of reagent, the developed colour is matched with the control solution and accordingly calculated. The colour changes from pink to yellow depending on the concentration of the flouride present in water samples.

Iodine

About 50 mg of iodine is present in a human body weighing about 70 kg. Probably all the cells of the body contain iodine, but 70 to 80% of total iodine content of the body is concentrated in the thyroid gland alone. When the dietary iodine intake falls below 10 mg/day throxine synthesis becomes, inadequate leading to endemic goitre. Persons suffering from goitre become mentally dull, lethargic and lack motor coordination capacity. Tetraglycine hydroperiodide or galabaline tablets or iodised salt are available in the market for compensating the deficiency of iodine in water and diet.

Estimation

The method for iodine determination begins with the addition of 1 ml of 5% NaCl solution along with 0.5 ml of 1% of arsenious oxide solution to each of the 10 ml of collected water samples. After thoroughly mixing the content, 1 ml of ceric ammonium sulphate solution (prepared by dissolving 13.38 g of 44 ml conc. H_2SO_4 and diluted to one litre) is added. Within 15 minutes, 1 ml of ferrous ammonium sulphate solution (prepared by dissolving 1.5 g in 100 ml distilled water with 4.6 ml of conc H_2SO_4) is added till the colour disappears. After one hour, the colour changes from red to yellow, depending on the concentration of iodine present. The control solutions are also run side by side under similar experimental conditions for comparison.

Nitrogen

The elements, nitrogen and phosphorous, are known as biostimulants. Nitrogen is an essential building block element in the synthesis of protein. Therefore, its data are of paramount importance in evaluating tne nutritional status of the water bodies. Four forms of nitrogen are present in the sewage and consequently in the receiving aquatic ecosystem. Freshly polluted system, especially by sewage contamination, shows higher concentration of ammonia nitrogen which in an aerobic environment changes to nitrites and then to nitrates. Barth (1953) calculated that, for the conversion of ammonia into nitrate, 4.5 mg oxygen for one mg ammonia oxidised, would be required. Nitrite nitrogen is relatively unimportant in water pollution studies, because it is unstable and easily oxidised into the nitrate form. Nitrate nitrogen is an indicator of past pollution in the process of stabilization and seldom exceeds 0.1 ppm in surface waters. Nitrate nitrogen as NO_3 should not exceed 45 ppm in drinking water, because of its serious and occasionally fatal effects on infant.

High concentration of ammonia nitrogen indicates that the pollution is fresh. With ageing, the concentration of albuminoid nitrogen goes on increasing while free ammonia concentration decreases in more or less the same proportion. 0.01 ppm of free NH_3 and 0.1 ppm of albuminoid NH_3 concentrations are the maximum allowable limits for water used for public consumption. Ammonia is one of the major constituents contributing to the toxicity of municipal waters. At high pH free NH_3 and its hydrolysis products are toxic in low concentration while at low pH, they are converted into ammonium ions (NH_4^{+1}) which are comparatively non-toxic.

Nitrate Nitrogen

The increasing application of fertilizers in agricultural lands results in water pollution due to the leaching of nutrients like N and P into the drains. The presence of these pollutants only in about 0.3 ppm of nitrate nitrogen concentration and 10 ppb soluble P in water, are needed to support the excessive production of algal blooms and aquatic weeds caussing undesirable colour, taste and odour, thereby degrading the aesthetic quality of surface water bodies. Further, it has been estimated that the nitrat nitrogen content varies from 0.04 to 5.0 ppm in the rivers of the world. A survey of the well waters in low lying areas around Patna, Bihar and elsewhere in India reveals that in certain wells, the nitrate-nitrogen was as high as 240 ppm which is considerably above the WHO limit of 10 ppm nitrate as nitrogen. Such high amount of nitrogen in well water is enough to produce disease known as methemogiobinemia in infants. 10 ppm of nitrate nitrogen concentrations interact under acidic condition with secondary amine compounds present in the water to produce nitrosoamines which are well known carcinogen. Another serious repercussions of high nitrate nitrogen concentration, is that non-pathogenic bacteria which live freely or in symbiotic association with plants, may get destroyed, resulting in a serious breakdown in the self-purification process of the aquatic ecosystem.

Procedure for Determining Free Ammonia and Albuminoid Ammonia Nitrogen Estimation

250 ml of water sample is usually distilled off in two different stages for measuring nitrogen in the above mentioned two forms. Accordingly, 250 ml of water simple is first distilled off at pH higher than 11.0. Such high pH value is achieved by adding 2 pellets of sodium hydroxide (NaOH). The distillation is continued, till 50 ml distillate is obtained (A). Thereafter 50 ml of alkaline potassium permanganate solution is added to the remaining constituents in the distilling flask and the distillation is continued till another 50 ml of distillate is collected in another nesselar tube (B). Both the distillates A and B, along with a series of control samples in separate nesselar tubes are arranged on white porcelain based stand. To each of the nesselar tubes, 2 ml nesselar reagent (K_2HgI_4) is added. Nesselar reagent is prepared by dissolving 35 g KI in 100 ml distilled water to which saturated solution of $HgCl_2$ is added. Subsequently, 120 g of NaOH is added, dissolved and diluted to 1 litre. Thereafter, the colour developed is matched with standard solutions. Supposing, if the concentration of ammonia is found to be 0.01 mg in the examined water sample after comparing it with the standard solution of ammonia, the sample examined is 250 ml therefore 1000 ml of water sample will show 0.01 × 4.0 ppm.

4-hours Oxygen Consumption Test

This test is used to measure the easily decomposable organic content present in water and waste simples. Mild oxidising agent like potassium permanaganate ($KMnO_4$) in acidic medium at 37°C is used to oxidise the organic materials. 125 ml of water sample is taken, to which 25 ml of N/80 potassium permanganate solution is added along with 10 ml of N/2 sulphuric acid. The whole content is kept at 37°C for 4 hours. At the end of the incubation period, one spoon of solid potassium iodide (KI) is added and titrated against N/40 sodium thiosulphate ($Na_2S_2O_3$) salution, using starch as indicator. The result is calculated as given below :

4-hours oxygen consumption value in ppm=

$$25(1\text{-}S/C) \times \frac{100}{\text{vol of sample taken}}$$

Where, S is the volume of titrant in ml with sample and C is the volume of titrant in ml consumed in control sample.

Bio chemical Oxygen Demand (BOD)

BOD represents the intensity of bio-degradable matter remaining in the stream at any time and DO (Dissolved oxygen) shows the ability of the stream to purify itself through bio-chemical process. BOD tests show the amount of molecular oxygen required by bacteria to reduce the carbonaceous materials. It is a bio-assay procedure that measures the oxygen consumed by living organism, while utilizing the organic matter. Bio-chemical oxidation is a slow process and therefore, 20 to 30 days are required for the complete degradation of the waste. Oxidation is achieved upto 95 to 99% in 5 days in case of glucose, about 65% in case of sewage, while 40 to 50% of the organic matters present in any other industrial waste water samples during the same incubation period. Both, the 20°C incubation temperature, used as an average value for slow moving water bodies in temperature climate and easily reproduceable in an incubator, and the 5 days incubation period, were chosen by British Royal Commission for sewage disposal because none of the rivers in England has a reaching time to the sea of more than 5 days.

Estimation

For the BOD test to be performed the sample of water from rivers are taken in duplicate without going for any dilution. In case of strong sewage or industrial wastes, the samples drawn are suitably diluted so that the DO present in the sample may not get completely exhausted during incubation period. Dilution is thoroughly aerated and enriched with sewage water nutrients for the microbes. 5 to 10 ml of per litre of diluted water is added as seeding material, prior to subjecting to the process of aeration. After suitably diluting the strong waste water sample in duplicate, one is put under incubator at 20°C for 5 days and the other set is put for DO determination. After the incubation period is over, DO contact in the sample is again estimated. The control sample is also taken to ensure the correct value for BOD. Finally BOD value is calculated as such :

BOD (ppm) = Depletion in DO (mg) × dilution factor where, the dilution factor is ;

$$\frac{\text{sample taken (ml)}}{\text{\% of dilution}}$$

Chemical Oxygen Demand (COD)

COD test shows the oxygen equivalent of the organic matter that can be oxidised by using a strong oxidising agents *e.g.* potassium dichromate in acidic solution, at elevated temperature, for two and half hour Silver sulphate is used as catalyst to oxidise the oxidation resistant organic compounds. In case where BOD test fails, COD succeeds. For example, when, municipal or industrial waste containing organic compounds that are toxic to biological life, produces erroneous BOD results. COD value is in general always higher than that of BOD The ratio of COD/BOD shows the types of treatment required for the particular waste water sample.

(i) When BOD : COD-0.6, it means that the waste is rich in putreascible matter. Therefore, the best treatment for such type of waste is biological one.

(ii) When BOD : COD is between 0.3 to 0.6, it means that the waste water requires acclimatization, prior to subjecting it for the biological treatment.

(iii) If and when BOD : COD = 0.3, the chemical treatment is the best for such waste.

Estimation

COD test is done by adding 20 ml of 0.25 N solution of potassium dicromate along with 75 ml of conc. sulpburic acid to

100 ml of waste water sample, in a digestion flask of 250 ml capacity. About 5 g of silver sulphate as catalyst is added to the reaction mixture. The whole set along with reaction mixture is refluxed for two hours. Thereafter, it is cooled under tap water and then titrated against 0.25 N solution of Mohr's salt (ferrous ammonium sulphate) using 2 to 3 drops of ferroin (prepared by dissolving 1.4 of 1 : 10 phenonthroline monohydiate together with 0.69 g of ferrous sulphate in 100 ml distilled water) solution as indicator, till the end point is reached. A blanck sample is also digested in similar experimental condition. The value of COD is computed in ppm as given below

$$\text{COD (ppm)} = \frac{(a-b)\, c \times 8000}{\text{volume of sample (ml)}}$$

where, a = vol. of titrant used (ml) in blank,

b = vol. of titrant used (ml) in sample

c = factor of the titrant

Phosphate as P

50 ml each of water sample along with a suitable control are taken in a series of nesselar tubes. To each of the tube, 2 ml of ammonium molybadate solution prepared by disolving 25.0g ammonium molybadate in 175 ml of distilled water (soln A). In another vessel, 280 ml of conc. H_2SO_4 is added to 400 ml distilled water (Soln. B). Solutions A and B are then mixed together. Thereafrer, five drops of freshly prepared stannous chloride solution of 2.5% strengh are added. The whole content of the tube is mixed, which result in the development of blue colour in water as well as in control sample. The visual colour comparison reveals the concentration of phosphate as (P) in ppm by multiplying with 20.

Silicate as SiO_2

Similar method as that for phosphate estimation is carried our for silicate estimation in water sample by adding 1.1 HCl and 2.0 ml of ammonium molybadate solution. The whole content is thoroughly mixed and kept for five to ten minutes. Thereafter, 1.5 ml of 10% oxalic acid is added. The yellow colour thus developed, is compared with those of control solutions and accordingly calculated by multiplication with 20. The result is expressed as SiO_2 in ppm

Nitrate as Nitrogen

To 50 ml of water sample, 2 ml of 30% sodium chl ride solution along with 10 ml conc H_2SO_4 are added and mixed gently. Brucine reagent (0.5 ml) is then added. The whole reaction mixture is put on water bath for 20 minutes, the colour developed is compared visually with suitable control solution and calculated by multiplying with 20. This gaves the valve for nitrate as nitrogen in ppm.

Sulphate as SO_4

100 ml of water sample with not more than 40 ppm of a sulphate is taken in nesselar tube. Similarly, a series of control samples is taken in different nesselar tubes. To each of the tube, 5 ml of conditioning reagent is prepared by disolving 76 g NaCl, 30 ml conc. HCl and 100 ml of 95% ethyl or isopropyl alcohol in 300 ml distilled water. After thoroughly mixing the whole content, oneospoon (say 200 mg) of barium chloride ($BaCl_2$) crystal is added. The turbidity developed is visually matched with that of the control solution of the sulphate within 5 minutes of the $BaCl_2$ addition.

BACTERIOLOGICAL PARAMETER

Most Probable Number MPN)

MPN count of water samples, drawn from the respective sources is determined by using multi-tubes dilution techniques. In the procedure, the McConkey lactose bile salt media are filled up in test tubes of different sizes and sterilized under 15 lbs pressure per square cm at 14 °C for 15 minutes. Thereafter, the tubes are arranged in four to five series, with three tubes in each series, for each sample of water to be examined. To each of this set, water sample is tranferred through sterilized pipettes. Each series contains different volume of original water samples, *i e.*, the three tubes of the first series are inoculated each with 10 ml the second series of three tubes with 1.0 ml each, while 0.1 ml each for the third and so on up to the fifth series by adding 0.01 ml and 0.001 ml respectively to all the three tubes of fourth and fifth series. Thereafter, all the sets are put at 37×2°C for 48+2 hours, A the end of the incubation period, each tube if examined for gas production in the durhms tube. The tubes having gas probuction are marked as positive ones and those without gas production are discarded. Numerical values for MPN count present in 100 ml of water sample are enumerated with the help of the already worked out chart, based on the formula developed by Thomas (1942). Mathematical expression for that formula is given below :

$$\text{Bacteriological count/100 ml} = \frac{\text{Number of positive tubes} \times 100}{\text{(ml in negative tubes)} \times \text{(ml in positive tubes)}}$$

Fecal Coliform Count (FC)

All the positive tubes of MPN count are sub-cultured in duplicate, in brilliant green lactose broth (BGLB) media. One set is put at 37°C for 48 hours and other set at 44×0.2°C for 24 hours. The values derived from the former set show the confirmatory number of MPN count/100 ml whereas the latter set shows the numerical values for the fecal coliform counts present in 100 ml of water sample under investigation.

Escherichia Coliform count (E. Coli)

All positive tubes of either fecal coliform or MPN are further sub-cultured in a series of three test tubes of McConkey bile-salt lactose broth and put for incubation at 37°C for 24 hours. Again, the gas production in each tube is examined. The positive tubes are allowed to proceed for further treatment as follow ;

The turbid inoculnm from each of the + ve tube is inoculated thorugh loop of standard size in the McConkey agar plate and put for incubation at 37°C for another 24 hours. At the end of the specified time, the well developed pinkish, flat and transparent colonies are further inoculated in peptone water and McConkey media The whole set is again put at 37°C for another 24 hours. The former, on producing positive reactions with Kovac's reagent and the latter, showing gas production, are utilized for the numerical assessment of escherichia coliform count, present in 100 ml of sample under investigation. Kovac's reagent is prepared by dissolving 5 g para dimethly amino benzaldehyde in 75 ml of either isoamyl alcohol or absolute alcohol and 150 ml of conc. H_2SO_4.

Fecal Streptococci Count (FS)

Fecal streptococci count is measured by apopting similar methods as for MPN and other counts, as discussed earlier and expressed per 100 ml of water samples. The only difference is the use of nonlactose media in the case of F. streptocci estimation. Therefore, azide dexirose broth (ADB) media are used in the tubes arranged in 4 series of three tubes each. To each of the tube, water sample in different dilution *i.e.* 1.0, 0.1 and 0.01 ml are inoculateh and put at 37°C for 48 hours of period. The tubes showing well marked turbidity are considered as positive tubes.

Further, the positive tubes are subcultured in ethyl violet azide (EVA) media and put to incubation for another, 48 hours at 37×2°C. The formation of ring at the bottom of the tube, confirms of fecal streptococci in the sample. The numerical values are ascertained again by using the Thomas chart, as used in evaluation of the MPN and E. coliform counts.

5

Techniques of Water Treatment

Introduction

A typical drinking water treatment plant is able to clarify the water and makes it hygienically safe. It may also add lime and sodium carbonate to make the water softer, and it may mix the water with activated carbon to improve the odor and the taste. The water is generally made to chlorinate at the and of treatment (sometimes at the beginning, too) with enough chlorine to kill essentially all bacteria and viruses and leave a residue of chlorine sufficient to keep the water safe until it reaches the water tap.

As water being treated for drinking purposes often is contaminated by organic substances, chlorination can produce some organic compounds of chlorine. Some that have been detected have been chloroform and carbon tetrachloride. These compounds are suspected of being carcinogens

No alternative technology to chlorination is immediately available. Ozone is used by special permission in one very small system. Ozone has been highly toxic to bacteria, and it destroys viruses more efficiently than chlorine. It leaves no taste. The techology for the use of ozone has been somewhat more expensive than for chlorination, and we have slight long-term experience in the use and effectiveness of ozone. Thus the government must move very carefully before requiring a switch from chlorination. Eventually the change could be made.

The methods of purifying waste domestic and industrial waters are largely traditional and have proved extremely effective in the past. However, in recent years further methods of purification have been introduced, particularly in the field of industrial water treatment.

Use of Coagulants

For effective removal of impurities from water it becomes essential that these have first flocculated, that, have been in a from that could be removed readily by filtration method. The aim has been to obtain particles in excess of 120 microns in diameter; these tend to settle readily in typical aqueous medium. Particles between about 3 and 100 microns tend to settle too slowly for practical purposes.

Flocculation seeks to achieve agglomeration of small particles into larger units and there by speed up the settling process. The most widely used flocculating agent has been alum, a complex salt of aluminium. When this has been added to an alkaline solution a voluminous aluminium hydroxide flocculate has been produced, which readly sinks to the bottom because it has been heavier than water. In its passage downwards it triaps tiny particles suspended in the water and thus removes them.

A recent development has been the use of a number synthetic andnatural and polymers. These form 'bridges' between adjacent solid particles and cause them to sedlment. In many instances polymer sedimentation has been not as effective as alum, because small particles, particularly those with a diameter below 2 microns, tend to possess a (usually negative) electric charge. Alum is able to dissipate this charge through the presence of positively charged aluminium ions in solution. However, metal salts have been not without their disadvantges. They are expensive, sometimes give inadequate flocculation and may also leave a 'carry-over' of fine turbidity.

Special synthetic polymers are being manufactured which are having the properties of bridging and discharge of electrical potential. Only water soluble monomer polymer systems can be used for waste water clarification. The following poly-electrolytes have been used.

(a) *Non-ionic types.* Polyols, polyethers, polyamides, polyvinyl heterocyclics;

(b) *Anionic groups.* Carboxylates, sulphonates, phosphonates;

(c) *Cationic groups.* Amines, quaternary ammonium compounds, sulphonium and phosphonium compounds.

The most common of all flocculants have been the polyacrylamides which are having molecular weights in the range 4-10 million.

Flash Distillation

Pure evaporation of sehwater and other brackish water has been still the most widely used of all methods of recover the pure water. The heat needed to boil a kilogram (litre) of pure water at 100°C has been nearly 2.27 MJ; but the heat needed to boil concentrated solutions of salts is greater as the presence of the salts elevates the boiling point. The main economy of the process has been dependent upon obtaining adequate heat economy.

This could be mainly achieved in two ways. First, solutions can be evaporated at lower pressure when the latent heat needed has been markedly reduced; and second, attempts can be made to use low grade waste heat for the process, especially using waste fuels like refuse, petroleum tail gases, etc., for operating seawater distillation plants There has been a considerable future in the operation of seawater desalination plants which use distillation on the total energy principle. The most commonly used process employing distillation as a basis has been flash distillation.

Flash distillation is involving a technique of allowing the seawater to boil successively in a large number of chambers, each of which has been maintained at a lower pressure than the one before, to match the lower temperature of the water. The method of flash distillation of seawater has been especially promising when related to the use of waste heat produced in nuclear power stations. It has been feasib e to design a nuclear power stations. It has been feasible to design a nuclear power station, operating on the total energy principle, to produce 400 MW of electrici y and 283,000 m^3 of freshwater per day.

Operating cycles of a flash evaporator. The flash evaporator is having twenty to fifty chambers is sequence, which are operating at successive lower pressure. When heated brine flows from one chamber to the next, some of it 'flashes' off into water vapour, which then condenses on colder condenser tubes and drops as distillate into trays to be led away into storage.

The brine, when passing from chamber to chamber, gets progressively cooler and ii has been this same brine which is eventually pumped back through the condenser tubes to act as the coolant in the condenser section of each chamber. It becomes progressively hotter as it does this; consequently, when it reaches the heat input section before re-entering the first flash chamber, its temperature need only be raised a few degrees. The heat would be narmally supplied by low-pressure steam which, in its turn, has been readily obtained by utilization of waste heat from primary power generating sources.

As can be seen from Fig. 5.1 the heated brine has been passing from the heat input section to the first flash chamber (1) and from there successively through all the chambers down to the coolest one (6), flashing off a certain amount of water vapour at each stage. It could be then extracted by the brine circulator pump and returned to the tubes of the heat recovery section at (4). In the last few stages—the heat rejection section (5) and (6)—cold crude sea water has been pumped through the tubes.

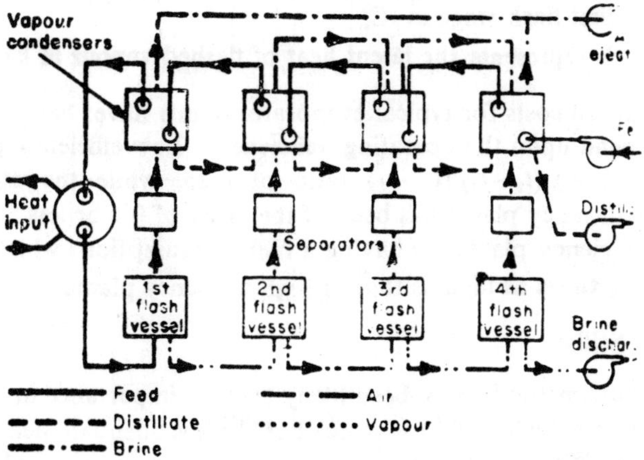

Fig. 5.1. Layout of a multi-effect flash installation using waste heat from diesel engines.

This allows condensation to occur in these stages and also extracts an amount of heat equal to that put into the evaporator at the heat input section, thereby allowing permitting a continuous cycle of operation. Some of the seawater, after chemical treatment to disallow scale formation, is added to the circulating brine to make up for the distillate extracted and for brine which must be discarded in order to keep the solute concentration in the evaporator within specific limits.

When seawater has been heated gases such as oxygen and carbon dioxide are given off. These could be extracted by a vacuum pump or ejector system.

The heat needed in a flash evaporator to produce unit weight of distillate has been given by the following equation :

$$(t_e - t_1)/(t_e - t_f) = L$$

where t_e represents the temperature of the water entering the flash stage in °C;

t_1 represents the temperature of the water leaving the tube system in °C;

t_f represents the temperature of the water leaving the first flash stage in °C;

L represents the latent heat of flashed vapour in kJ/kg.

Capital costs for typical evaporator plants have been found to depend upon the operating efficiency. Low efficiency plants are having a $(t_e - t_1)/(t_e - t_f)$ ratio of 0.25, while the ratio for high-efficiency plants has been of the order of 0.1 or less. Thus low efficiency plants are having a fuel consumption which has been 2½ times as high as that of high efficiency plants.

Solar Stills

These methods have been likely to gain importance in areas of the world where fuel is scarce but there is adequate heat from the sun. Solar radiation energy available in many parts of India and Pakistan has been of the order of 20 MJ per m² land surface area per day for some five months in the year, and even

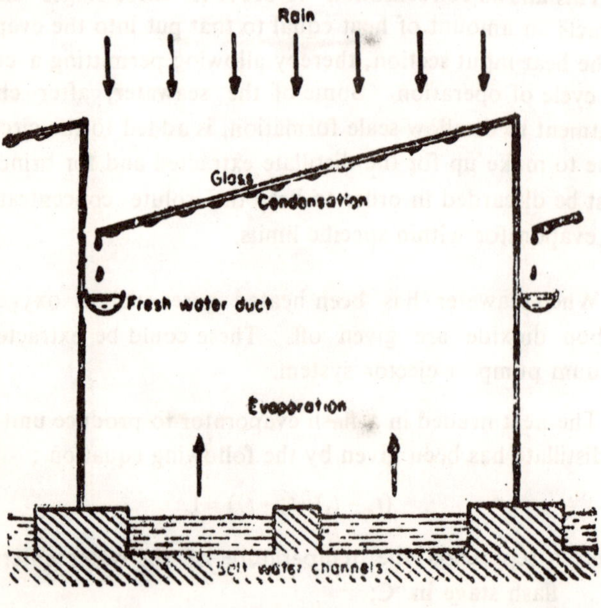

Fig. 5.2. Use of solar energy for desalination of salt water, The large 'greenhouses' are also used for the collection of rain water

in mid-winter the solar heat has been still more than 10 MJ/m per pay.

A system developed in India uses a large area of glass positioned between 30 and 60 cm above the surface painted black, along which seawater is allowed to flow. Solar heat makes water to evaporate, leaving the salt behind, and condensation takes place on the glass cover. The water drips off to an aluminium channel and runs to a storage vessel. The unit has been also equipped to collect rainwater. The cost of getting water from the experimental 2 m^3 per day stills, which have been able to extract around 3-4 litres of distilled water from each m^2 of still surface erected, is about 18p per m^3. It is now thought that with larger installations it becomes possible to use this system anywhere the present cost of water exceeds 70p per 1000 U.S. gallons or 18.5p per m^3. Such conditions have been found in many parts of India and other arid regions of the earth.

A large solar distillation plant was built near Daytona Beach, Florida. The feed water is having 3 2 per cent dissolved salts and the area of land covered is 220 m^2. Each unit is consisting of grooved concrete curbs laid 1 m apart to form a bay 20 m in length.

There have been plastic basin liners and clear plastic cover domes. Underneath the brine has been a thick layer of insulation, so that the heat generated inside the still has been not conducted away to the soil. Water evaporates from the feed water and condenses against the plastic dome cover, running down towards galvanized steel distillate troughs at the sides of the stills. Solar stills, based on similar designs to the above, are also erected in Greece and Spain.

For a basin type solar distillation plant, with a capacity of 180 m^3 of water per day, covering a ground area of 50,000 m^3 plant investment cost has been about £ 2,00,000; as far as running costs are concerned, amortization accounts for 70 per cent of the daily costs.

The next major items of expenditure have been taxes, insurance and interest on working capital, which add a further 18 per cent. The final cost of water supplied works out at 35p per m^4 very much higher than with distillation systems which use purchased fuels.

However anomalous it may be that where the fuel has been free the costs has been higher, it has been certain that solar stills will not be economic in comparison with other methods of water desalination for many years to come.

Ion exchange

The cations commonly present in natural waters including seawater, have been calcium, sodium, potassium, magnesium, iron, and traces of other metals like manganese, aluminium, copper and zinc. The usual anions have been bicarbonate, chloride, sulphate and traces of nitrate and fluoride.

Ion exchange purification takes place by passing water first through an exchanger where cations get exchanged for hydrogen ions, followed by passage through an anion exchanger bed where acid radicles have been replaced by hydroxyl ions so that :

H^+(hydrogen ions) + OH^- (hydroxyl ions) → H_2O (water).

Naturally, it becomes necessary to reclaim both the cation and the anion exchange resins as they become spent. This is normally carried out by reacting the former with strong acid and the latter with custic soda. There have been a number of more sophisticated plants including so-called 'mixed bed' de-ionizers where special methods of fluidization are able to separate the different ion exchange resins to enable them to be reclaimed separately. In the past it was regarded uneconomic to use ion-exchange plants for the desalination of water with a solids content in content in excess of 1.5 per litre.

By employing a cascade system of desalination it now becomes possible to treat water with a salt content well in excess of 10 g per litre. Some Russian plants have been using ion-exchange membranes for the desalination of waters having up to 15 g of salt per litre. The technique used has been a combination of electrodialysis, and ion exchange using membranes which are referred to simply as MK - 40 and MA - 40. These are having as surfare electrical resistance of about 12 ohm/cm^2, measure 150 cm × 50 cm have been able to exchange between 2.3 and 4.4 mg equivalent of ions per hour. The total electric power consumption needed has been 9 kWh per m^3 of drinking water produced.

A new American technique employs weak electrolyte ion exchangers to change sodium chloride into sodium bicarbonate first followed by removal of the cation (sodium) and then recovery of the carbon dioxide which is still present in solution. It has been reported that with this technique really concentrated solutions of salt could be treated efficiently and economically. The 'Sirotherm' process has been an ion-exchange process in which regeneration of the resin has been by heat and not chemicals. At present this process has been still at the laboratory stage, but there have been indications that it has been technically, feas ble to treat water with a salinity of up to 3 g per litre. It is thought that some other process may be required to reduce the salt content of seawater from its usual 35 g per litre to that at which an ion exchange process is feasible. This implies a combination of electrodialysis or reverse osmosis with an inexpensive ion-exchange technique like 'Sirotherm'.

Reverse Osmosis

When we keep a solution on one side of a semi-permeable membrane and pure solvent on the other, the solvent will pass through this membrane to dilute the solution The pressure which gets exerted by the solvent as it passes through is called the osmotic pressure.

Indeed, it is to be argued that in solution a dissolved salt behaves like a gas and obeys the gas laws. This implies that one gramme mole of cation or anion dissolved in 22.4.4 litres of water gives an osmotic pressure of exactly 1.013 bars (1 atmosphere) at a temperature of 0°C. Reverse osmosis has been found to depend upon the use of a somi-permeable membrane, which is capable of being traversed only by water and not by salt, combined with pressurizing of the solution so that water passes through the membrane, but the salt stays behind.

Due to the high pressures involved, the main difficulty has been in the making of a semi-permeable membrane of adequate strength. In general, membranes could be supported by a special carrier. The following semi-permeable membrane materials are tried, with favourable results :

(a) Cellulose acetate dissolved in glacial acetic acid together having small quantities of additives like magnesium perchlorate before being allowed to solidify on the carrier;

(b) A cellulose acetate/formamide solution in acetone;

(c) Ethyl cellulose mixes.

The working pressure on one side of the reverse osmosis cell has been normally of the order of 100 bars, while the other has been at atmospheric pressure. Supporting the semi-permeable membrane to avoid breakdown has been therefore a problem of considerable magnitude.

The first reverse osmosis plants was using perforated stainless steel plates and tubes as supports, but these have been reported to be too expensive for commercial plants. The best material found which was having the requisite strength and at the same time was not too expensive was fibreglass cloth laminate about 3 mm thick; this is having a compressive strength of 15 MN per m^2 and a tensile strength in excess of 150 MN/m^2. As far as tubular support materials are concerned, phenolic resins having a wall thickness of 2.5 mm seemed to provide some of the best results at the lowest cost.

The desalination cell has been so designed as to bring the saline solution at high pressure in contact with the semi-permeable membrane, with either plates or tubes to support the

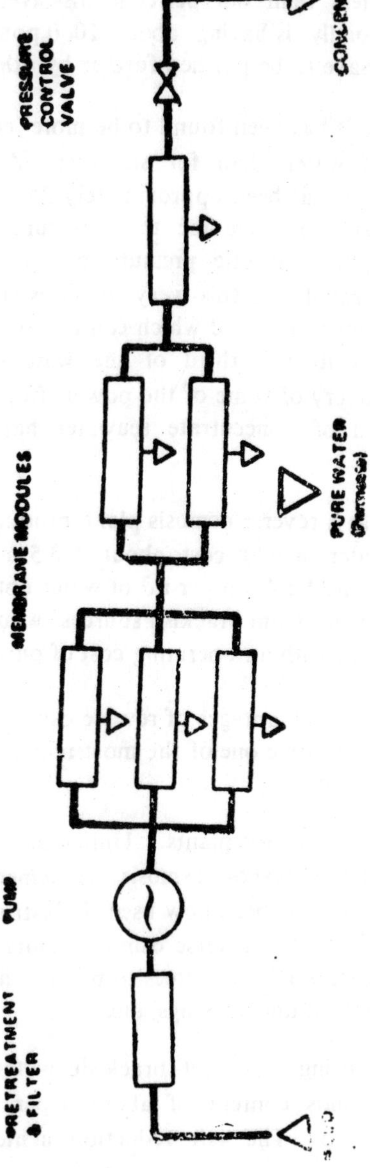

Fig. 4.3. Reverse osmosis flow sheet

membrane itself. As osmotic pressure tends to increase with increasing salinity, the through put rate for a given membrane is less with seawater (3.5 per cent dissolved solids) than typical brackish water (less than 0.5 per cent dissolved solids). The final product usually is having about 2000 ppm of dissolved solids, and may have to be purified further by other means.

Reverse osmosis has been found to be more economical for low salt content water than for seawater. As the osmotic pressure of seawater has been approximately 25 bars, it has been obviously necessary to overcome this pressure before reverse osmosis can begin. Osmotic pressure builds up as the feed stream is concentrated and this very much is able to limit the percentage of water in the feed which can be usefully recovered as freshwater; only about a third of the water can, in fact, be recovered. Recovery of some of the power from the flow of the waste stream of concentrate seawater has been usually attempted.

It is found that a reverse osmosis plant producing 40 m^3 of water from seawater would cost about £ 3.5 million, and its operating cost would be 4.5 p per m^3 of water obtained. A plant to recover pure water from brackish sources would be costing about £ 1.1 million with an operating cost of only 2.3p per m^3.

In spite of the disadvantages of reverse osmosis, this system has been likely to become one of the most important in the field of water desalination.

Practical rverese osmosis plants. Unlike many of the other desalination processes, reverse osmosis has emerged from the pilot plant stage and has been now used industrially. Paterson Candy International make reverse osmosis plants able to supply drinking water from brackish sources to private houses, hotels, farms, small industrial undertakings, etc.

The units are using a feed of brackish water with a maximum dissolved solids content of about 1 per cent and pH between 3 and 7.5. The salt reduction achieved has been between 85 and 95 per cent, depending upon the nature of the ions. The semi-permeable membrane in Paterson Candy units has been modified cellulose acetate cast on the inside of disposable paper/polyster tubes.

Normally membranes can find use continuously for between 12 and 24 months before replacement becomes necessary. An electric motor has been used to produce a pressure of 40 80 bars, depending upon the nature of the feed. Power consumed is, on average, about 2.5 kWh/m^3 of desalinated water—which compares favourably with 12 kWh/m^3 for distillation and 7 kWh/m^3 for freezing processes.

A plant measuring $2.6 + 1.5 \times 1.3$ m and weighing under 2 tonnes has been able to produce 45 m^3 of desalinated water per day. This has been incidentally, the largest plant currently made by this firm, which produces equipment having capacities from 2.3 to 45 m^3 day.

Electrodialysis

In electrodialysis, as in reverse osmosis, a semi-permeable membrane permits pure water to pass, but stops salts. Whereas in reverse osmosis the passage of water through the membrane has been induced by physical pressure acting on the saline solution, in electrodialysis this work has been carried out by an electric current.

The basic idea has been that the saline solution has been enclosed by membranes and an electric current induces cations (metals) to migrate outward towards a cathode, and anions (the non-metallic ions) to migrate towards an anode, thereby progressively desalinating the solution.

Unfortunately, the procedure in practice has been not quite as simple as this. For progressive desalination to continue, the energy exerted during the electrodialytic process must also overcome osmotic pressure which is able to force clean water back into the solution.

As in reverse osmosis, membranes have to be very tough indeed for withstanding the high pressures applied. At the present time electrodialysis has been most widely used in the Soviet Union, and the most common membrane has been polyethylene backed by nylon fibre.

In general the greater the current density used in any given dialysis cell, the smaller the area of membrane required. But increased current density also necessitates increased flow rate to prevent the membrane from being polarized. Optimum current density has been found to vary with the concentration of salts in water to be treated, the cost of the membrane and the cost of the power expended.

For seawater having a salt content of 3.5 per cent and a power cost per kWh amounting to 0.14 per cent of the cost of membrane per m^2, the optimum current density has been 1.06 A/dm^2. For other conditions, the optimum current density may decrease between 0.7 and 3 A/dm^2.

Electrodialysis desalination finds use in the U S S.R mainly for water supply in sparsely populated areas, farms and watering places at distant pastures - as well as for expeditions venturing to areas where there has been no fresh water but there has been access to underground or surface salt water. Although the process has been basically an expensive one - both from the point of view of initial cost and of power needed - it is having the considerable advantage that it can be completely automated, needing no resident labour.

In Western countries electrodialysis for seawater desalination has been not generally regarded economically viable compared with other methods, mainly due to the large quantities of power needed. However, if at some future date the cost of electric power must fall appreciably, electrodialysis may become competitive.

Separators of expanded rigid PVC have to be used to support the membranes and to create turbulence in the stream. Two plants have been built, one with a daily output of 77 m^3 and the other with a daily output of 151 m^3, using membrane areas of 12 and 40 m^2 respectively.

The biggest American electrodialysis plant is the 2460 m^3 per day plant at Buckeye, Arizona. This brings down the salinity of the water from approximately 2.1 per cent to 0.5 per cent at a cost of about 5p per m^3 of water treated. Electrodialysis provides the most economic results when the installed plant capacity is large and when the feedwater salinity is comparatively low.

Ion Exchange Techniques

There are four basic ion exchange resins which have been able to absorb even small quantities of dissolved cations from solutions. These have been sulphonic-phenolic resins, sulphonated coal, carboxylic resins and sulphonated polystyrene. For all of them the reaction which occurs can be expressed as :

$$H\bar{R} + M^+ \rightarrow M\bar{R} + H^+$$

where M^+ represents the metallic ion (its valency does not have to be one) and $\bar{R}$ is the net-like resin structure. Once the ion exchange resin gets absorbed its capacity of metallic ions, it could be reclaimed by means of a strong acid :

$$M\bar{R} + HCl \rightarrow H\bar{R} + MCl$$

The metal can then be able to recover from the concentrated chloride or other salt. Ion exchange of this type has been found to be very effective for many metals especially, lead, copper, zinc, mercury, chromium and nickel. For arsenic an anion exchange resin is able to remove the metal in the form of arsenates and arsenites; reduction of arsenic contents of waste liquors to below 5 ppm can be attained without difficulty.

Copper and zinc can be removed from waste liquors so effectively that a final effluent is having less than 0.001 ppm of either metal. It has been vitally necessary, as it has been shown that even concentrations of 0.02 ppm of these metals are having very harmful effects on fish.

Ion exchange techniques must be used, where possible, in conjunction with precipitation methods, because they have been most successful when the initial concentration of dissolved metal ions has been already fairly low.

Self-purification of Waters

In the various zones of receiving streams (surface waters such as streams, rivers and lakes into which unpurified or purified waste waters are channeled and by which they are carried away), living creatures are to be found which are able to utilize any nutrients that are available. Natural self-purification of the waters by the decomposition of putrefactive substances, after a given period and distance of flow of the receiving streams, is due to these creatures. Under favorable conditions the pollution of the waters is substantially corrected in this way, so that the original pure state is at least approximately restored.

The self-purification processes depend mainly on biological processes (self-purification by organisms) or chemical processes (primarily oxidation and reduction processes), the effectiveness of which is determined by physical factors such as velocity of flow, state of the water bed, ratio of foul to clean water, temperature and depth of water, intensity and duration of solar radiation and fineness of waste substances. The main burden of self-purification falls on the biological processes.

The organisms contributing to the process of self-purification include fungi, algae, bacteria, protozoa, crustaceans, shellfish, worms, insect larvae, snails fish, big predatory animals and waterfowl.

In the metabolism of the organisms these substances are incorporated into the body matter or ultimately broken down into water and ca bon dioxide for the purpose of providing energy. In order to sustain these processes, dissolved oxygen is absorbed from the water. Owing to their autotrophic manner of feeding, the algae containing chlorophyll are in a position by day to return the oxygen so absorbed. If the water is overrich in nutrients, *i.e.*, overloaded with plant nutrients such as nitrogen or phosphorus, there can well be an explosive proliferation of algae. This will result in a shortage of oxygen as, owing to the limited absorptive capacity

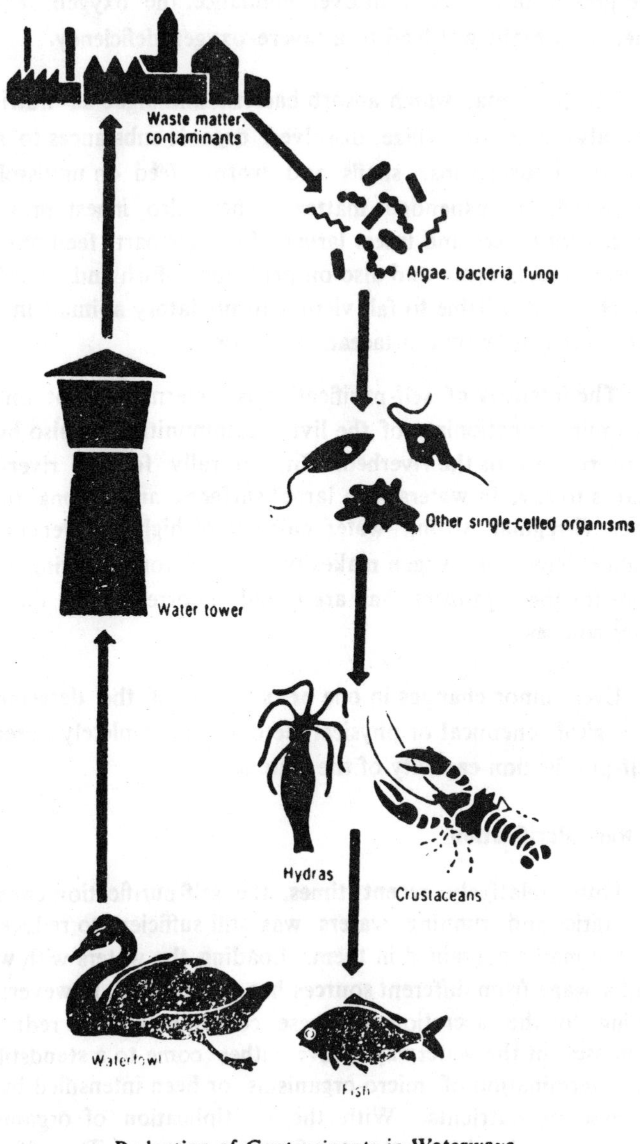

Reduction of Contaminants in Waterways

depends largely on the water temperature. At higher temperature the saturation point is reached more rapidly. If algae are present in the water in overabundance, the oxygen requirements at night will lead to a severe oxygen deficiency.

The protozoa, which absorb bacteria and algae as nutrients, are also able to utilize dissolved organic substances to some extent. Crustaceans, snails and worms feed on undissolved, deposited or suspended matter. They also ingest protozoa. Bigger crustances and insect larvae, for their part, feed on small crustaceans, worms and also on protozoa. Fish and waterfowl, which are also liable to fall victims to predatory animals in their turn, often prey on crustaceans and worms.

The intensity of self-purification is determined not only by optimum functioning of the living communities but also by the conformation of the riverbed. In naturally formed river-beds that is to say, in waters with large surfaces and strong turbulence (irregular strong water currents of high turbulence), the higher input of oxygen makes for more favorable living conditions for the organisms than are found in corrected or dammed river courses.

Even minor changes in one or another of the determining biological, chemical or physical factors can completely upset the self-purification capacity of the waters.

Sewage Purification

Until relatively recent times, the self-purification capacity of static and running waters was still sufficient to reduce the contaminants contained in them. Loading the waters with waste and sewage from different sources has proliferated, however; and owing to the accretion of these contaminants, the reduction processes in the waterways have either come to a standstill by the extermination of micro-organisms or been intensified by the increase in nutrients. With the multiplication of organisms, more oxygen has been extracted from the waters. The self-purification capa- city of the waters has been overloaded, and the regenerative powers of the waters have been disrupted. The immediate consequences of such pollution have been the extermination of fish life, deposition of mud, and putrefaction processes.

The most serious loading of the waters stems from domestic and industrial waste disposal. The sewage is conducted through the drainage system to a purification plant, which as a rule operates in two stages : mechanical and biological treatment stages. The mechanical system comprises a grid, a sand trap and the preliminary settler. The coarser elements in the sewage are separated and retained by the gird. As the flow speed of the water slows, smaller mineral elements settle in the sand trap; the other removable matter is deposited in the preliminary settler, and the floating elements are removed by means of a scoop. The purification performance at the mechanical stage stands at about 20%–30%.

The water thus purified passes finally to the biological staget, which comprises a restoration basin (percolating filter) and a second filter, or final sedimentation tank. In the restoration basin, bacteria and other micro-organisms convert the suspended and dissolved matter into removable sludge by absorbing the particles and dissolved matter as a nutritive substratum and by forming cell lumps as they proliferate, which, owing to their increase in weight, sink to the bottom. The whole process is facilitated and intensified by artificial induction of air by means of bellows, rotors and cylinder pumps.

In order to ensure that the micro-organisms required for decomposition are present in sufficient quantities, parts of the micro-organism sludge deposited on the floor of the secondary filter tank are returned to the restoration basin. The purified water on the surface of the secondary filter tank is fed into the receiving stream.

The sludge from the primary and secondary settlers is digested in septic towers with the aid of anaerobic bacteria. By this process, methane gas is formed which can be used to provide energy for the plant. Previously the digested sludge was usually dried out in beds to compact it. In uew plants the preference is for mechanical dehyderation plants, such as centrifuges and presses, because of the smaller space they take up. The dehydrated sludge can then be further processed together with household refuse.

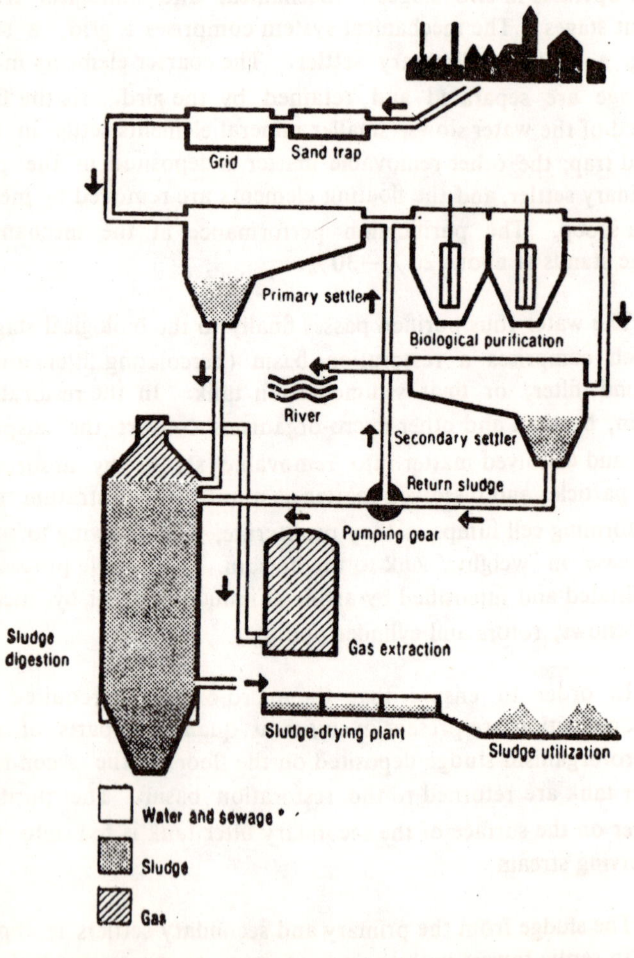

Plan of a Mechanical-biological Sewage Treatment Plant

Under optimum conditions the mechanical-biological sludge drying plant can achieve a purification performance of over 90%. For further purposes, however, this performance level is still too low, so that a further stage, the chemical stage, needs to be added. The chemical precipitation process employed serves to eliminate phosphates and other (especially industrial) pollution. With the aid to microsieves and similar technical appliances, organic residues can be further reduced. In addition, a hygienic improvement in the sewage may be achieved by means of chlorination, radiation, heating or ozonizing.

The Corrugated Plant Filter

For the purification of contaminated waste waters, a plant with corrugated-plate-filter muck piles can be installed wherever the water is charged with floating or deposited matter. The oil filters can be installed at any point in the petrochemical or mineral oil processing industry, in tanker depots, at airports, in big garages, etc. Plant for the removal of solids is installed for the purification of filter return water and for sludge filtration (*e.g.*, in the foodstuff and textile industries).

The corrugated plate filter depends for its operation on the force of gravity. In the filter the particles of contaminants do not need to rise the whole way to the surface of the basin, nor to sink to its floor, since the filter contains a number of corrugated plates, one above the other, representing in effect a series of surfaces and of floors. Any one particle need to cover only the distance between two plates and is then effectively deposited. The corrugations on the plates make for closer pooling of the contaminants from the water for onward transmission to the filter channels. Additional collecting channels on the inlet and outlet sides of the corrugated plate will ensure that the filtered matter is collected and carried away without being diverted (by the inlet or outlet currents) into the sewage water again or contaminating the clean water.

According to the specific gravity of the contaminants to be eliminated, the filter will be fed from above downward (for matter that is lighter than water, such as oil) or from below upward (in the case of matter than is heavier than water).

The way in which the filter plant and the filter itself function is illustrated in the following example, relating to oil filtering. The inlet tank produces an even distribution of the incoming contaminated water between the individual parallel cells of the filter tank, so that all the plate packs are equally loaded. In front of each pack a flow filter spreads the water flow evenly over the whole field. The oil and water mixture makes its way from above into the pack, which is set at an angle of 45° to the vertical. It flows down diagonally between the corrugated plates set one above the other. The flow between the plates is laminar; the oil droplets rise unhindered to the underside of the plate above, collect in the crests of the plate, and flow back upward to

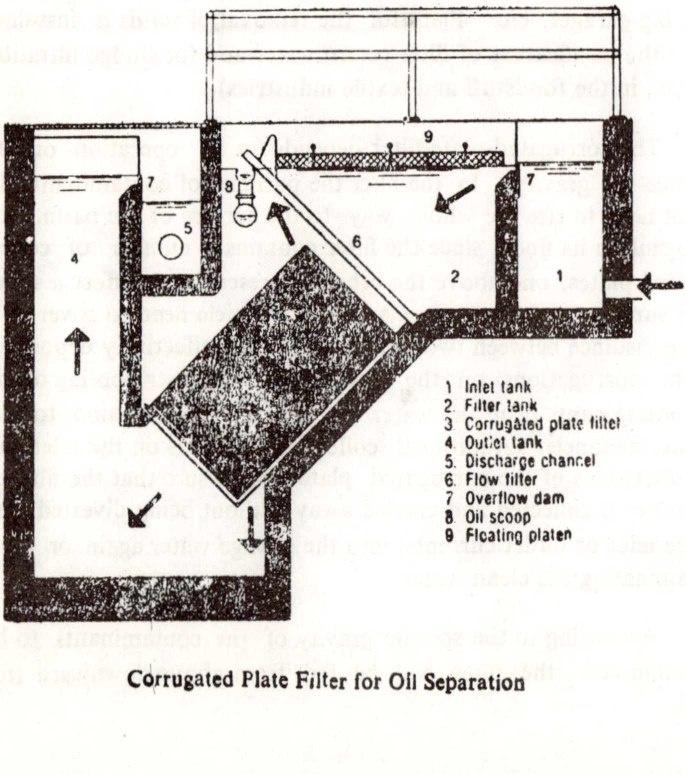

Corrugated Plate Filter for Oil Separation

the entrance of the plate pack. There the oil is caught up in the mainstream and carried away from the flow of water coming in to the surface of the basin. The sludge present in the water collects in the troughs of the undulations, sinks down with the current flow of the water to the outlet of the pack, where it is similarly taken up by the collecting channel, and is carried in the mainstream to the floor of the basin or into an elutriating funnel. The cleansed water passes into the outlet tank, and from there over an overflow dam into the outlet channel. The separated oil collects as a slick on the surface of the water in the filter tank and flows over an oil scoop or skin pipe into the sump. The filter tank is covered with floating plates of plastic form in order to reduce evaporation and offensive fumes.

For the separation of solid matter the direction of flow is reversed, but the way in which the entire plant works is, in other respects, the same.

The average performance of a set of plates, with a distance between the plates of 19 mm, is 30 cubic meters per hour. In special cases this average performance is reduced, however, to enable particles with lower rates of rise or fall to be thoroughly separated.

Composting of Sewage Sludge : The Counterflow Process

The counterflow process, which was developed for the composting of raw sludge from local sewage plants, is suitable for use in all sewage plant systems. The reactors are available in sizes of 25 cu m, 50 cu m, 100 cu m, and up to 500 cu m. Anaerobic or aerobic sludge stabilization (digestion tower) is superfluous. What is necessary, however, is drainage of the sewage sludge by mechanical means. For composting the sludge, owing to its high nitrogen and water content, an organic carbon carrier, such as peat, sawdust, chopped straw or lignite, is also required.

The sludge, fed into the sewage plant with a water content of about 98%, is dehydrated mechanically by means of presses and centrifugal driers down to a water content of 75%-80%. The dehydrated sludge can be mixed with the required organic carbon carrier by means of a mixing screw in the ratio of 1 : 1. In the case of a moving operation, instead of the carbon carrier, backflow (reflux) matter can be used in the proportions of 50% sludge, 40% reflux and 10% carbon carrier. The mixed material is raised by a sloping conveyor into the bioreactor, which consists of combined metal elements. In order to prevent heat dissipation, the reactor is insulated over its entire outer wall. The material provided passes slowly and continuously through the reactor from top to bottom (period of passage about 10 days). Ventilation is effected at the base of the reactor. The air injected through nozzles passes upward through the material. Control of the airflow is possible. Air is constantly removed at three points in the material store (reactor filling) and its carbon dioxide content examined with analysis equipment. Resistance pyrometers constantly check the temperature at six different points, which is then recorded on a six-point inkwriter.

The rapid group system is adopted for utilizing clean sewage sludge. By a regular supply of oxygen, optimum living conditions are provided for the reducing and converting micro-organisms already present in the sludge. In this way they are stimulated to exercise higher metabolic activity and faster germination. The air injected has an initial oxygen content of about 21%. It penetrates the material slowly by a counterflow process in an upward direction. This promotes the respiration of oxygen by the organisms. Oxidation processes thus get under way and the oxygen content of the air is reduced from below upward. In the upper field of the reactor the content of oxygen in the air falls to 7%-10%. In accordance with the airflow the temperature rises from below upward. In the upper region immediately below the condenser water zone the temperature builds up to levels of 75°-82°C (167°-180°F). The material, which is slowly and continually traversed by the reactor from top to bottom, must necessarily therefore pass through this heat accumulation zone so that hygienic treatment of the sewage sludge is assured. The end product, considerably reduced in volume and weight compared with the original material, is a hygienically flawless, friable material.

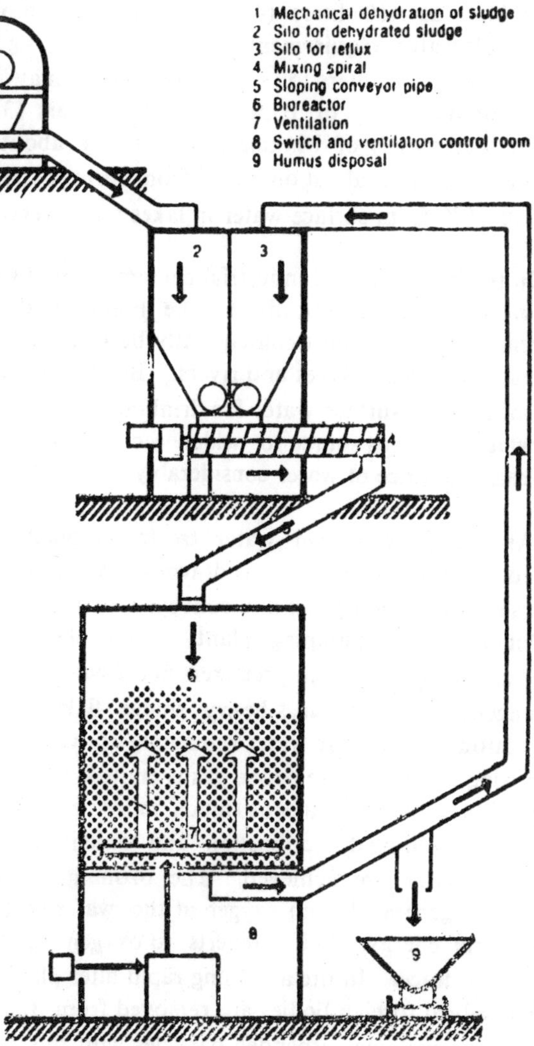

Sewage-sludge Disposal Plant

Water Supply : Preparation of Surface Water

Water consumption in big cities amounts to some 200-300 liters (210-315 quarts) per person per day, rising in peak periods to as high as 450 liters (472 quarts). On top of this come industrial water requirements. To produce 1 kg (1.2 lb) of plastics, up to 500 liters (525 quarts) of water may be needed, and to produce 1 kg of paper up to 3000 liters (3150 quarts. Of the total supply of drinking water today about half comes from groundwater, about one-third from springwater, and about 15% or 16% from surface water in lakes and rivers.

In thickly populated industrial centers, groundwater supplies ase now utilized practically to the limits of their capacity. Increasing recourse must consequently be had in the future to the surface water in lakes and rivers. But because of the high cost of treating surface water for drinking, the supply of drinking water from this source to wider sectors of the population will raise the price of water considerably.

An example of a river water treatment plant will be described : The untreated water is drawn from the river by means of a suitable suction plant and conveyed to the waterworks by a remote-control pumping plant. There the river water is sterilized with chlorine in premixers and treated with a flocculation agent. This produces heavy sludge flakes which sink to the bottom. The clear water remaining above flows on to the ozonizing plant. There ozone gas is emitted into the water in fine bubbles, by which means (much more intensively than by chlorine) additional sterilization and oxidation of dissolved organic matter are achieved. The ozone gas needed for this purpose is generated from oxygen at the waterworks by means of electric energy, and it reverts to oxygen again afterward, leaving no trace. In the adjoining rapid filter plant all the other still-dissolved foreign bodies are removed from the water, first of all the remaining suspended particles with the aid of a 2-3 meter thick quartz sand layer and then all the other still-dissolved foreign bodies by means of the active carbon filter inserted below it. The treated water is retained until its introduction into the filtrate reservoir.

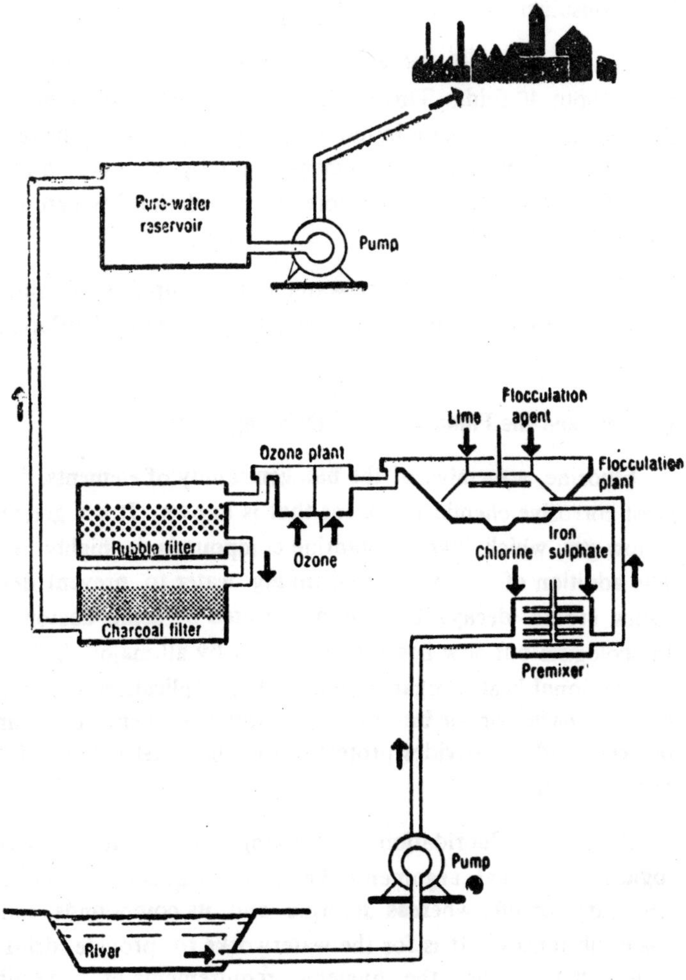

Plan of a Waterworks

A further possibility is to allow the treated water to seep away into suitable gravel deposits in the river valleys, to lower its temperature and assimilate it in other respects to natural groundwater, and after a specified period to remove it in the same way as groundwater together with natural groundwater. A pure water reservoir is needed to retain the water removed, from which the drinking water is conveyed by pumps to the final consumer.

Consumption of drinking water has risen in the past 50 years about 40-fold. This is explained, for the most part, by the enormous expansion in industrial production, by increased awareness of hygienic requirements, and by changes in man's consumption habits. Furthermore the population has practically doubled in the period.

The proportion of surface water in total supplies of drinking water is likely to expand in line with the constant development of industry.

Fluorine and the Fluoridation of Drinking Water

Fluorine, a member of the halogen family of elements, is the most corrosive chemical element that is known. It is a greenish-yellow gas, which, like the fluorine compounds, is highly toxic. The addition of fluorides to drinking water to prevent dental caries (tooth decay) had been opposed by some doctors and toxicologists but now has been endorsed by all major U.S. and international health organizations. The application of fluorides in tooth-pastes or for internal consumption hardens the enamel on teeth and so provides protection against destruction of the teeth by caries.

Against the fluoridation of drinking water, various toxicological and legal arguments have been adduced. Water is necessary for life, whereas fluorine and its compounds can be toxic substances. It is for the waterworks to provide drinking water that satisfies the hygienic requirements of comestibles (food). Fluoride at high concentration produces a corrosive effect on glass, steel and a number of other metals. Drinking water is treated with 1 mg of fluoride per liter of water. At concentrations over 2 mg per liter, people begin to react gradually with symptoms of mild dental fluorosis, as part of the daily intake of fluorine accumulates in the body. The fluorosis manifests itself as white flecks in the dental enamel. Pitting of the enamel can result from fluorine concentrations of 3-4 mg per liter. Only in a few places in the United States is the concentration above 4 mg per liter. In those areas the Environmental Protection Agency is working to reduce the fluorine level.

People do not absorb fluorine into their system only as an additive in drinking water, but fluorine is present in their diet in natural and artificial forms. This supply of fluorine in the daily diet varies in different countries and for different types of food; so that the total intake of fluorine in drinking water and foodstuffs cannot be assessed precisely. A further factor is the loading of the atmosphere with the fluorine waste gases from aluminum works and similar industrial plants, which can also result in an uncontrollable intake of fluorine.

Although fluorine has been shown active in a dental context only for children under 15 years of age, the entire water supply system is treated with fluorides. It is estimated that about 1000 liters of water would have to be treated with fluoride to enable a child whose teeth are still developing to drink 1 liter of treated water.

A difficult legal problem now arises from the fact that the consumer can in no way avoid being supplied with fluoridated drinking water. The situation amounts to enforced medication. However, the legal validity of fluoridation has been upheld in courts in the United States and in several other countries.

In many cities in the world, however, the fluoridation of drinking water introduced in the past on a trial basis has been abandoned. In 1971 the model project introduced in 1952 in Kassel-Wahlershausen, Germany, was suspended. In 1972 the fluoridation of drinking water in Sweden was discontinued. In the United States more than half the population drinks fluoridated water and that proportion is increasingly slowly.

Caries is not a disease caused by fluorine deficiency but is the result of bacterial action exacerbated by general nutrient deficiencies and lack of dental care. Fluorine medication can, at best, influence the symptom of caries, but not the disease itself.

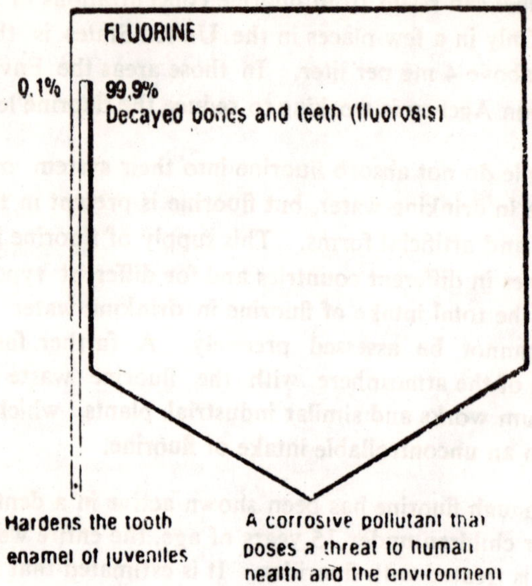

Effects of Fluoridation of Drinking Water

6

Water Supply Management

Introduction

Water is a chemical compound and may occur in a liquid form or in a gaseous form. All these three forms of water are extremely useful to man, providing the luxuries and comforts, in addition to fulfilling his basic necessities of life. Everyone of us knows how important and precious the water is. Whenever there is no water in our taps, we become helpless. No life can exist without water since water is as essential for life as air is. It has been estimated that two-third of human body is constituted of water. Water is absolutely essential not only for survival of human beings, but also for animals, plants and all other living beings. Further, it is necessary that the water required for their needs must be good and it should not contain unwanted impurities or harmful chemical compounds or bacterias in it. Therefore, in order to ensure the availability of sufficient quantity of good quality water, it becomes almost imperative in a modern society, to plan and build suitable water supply schemes which may provide potable water to the various sections of community in accordance with their demands and requirements. The provision of such a scheme shall ensure a constant and a reliable water supply to that section of the people for which it has been designated. Such a scheme shall not only help in supplying safe wholesome water to the people for drinking, cooking, bathing, washing, etc. so as to keep the disease away and thereby promoting better health; but would also help in supplying water for fountains, gardens, etc. and thus helping in maintaining better sanitation and beautification of surroundings, thereby reducing environmental pollution. Besides promoting overall hygiene and public health, it shall ensure a safety against fire by supplying sufficient quantity of water to extinguish it. The existence of such a water supply scheme shall further help in attracting industries (since industries require large amounts of water) and thereby helping in industrialisation and modernisation of the society, consequently reducing unemployment and ensuring better living standards. Such schemes shall, therefore, help in promoting wealth and welfare of the entire humanity as a whole.

Demand of Water

India is likely to have a population of around one thousand million by the end of the century, out of which the urban population will be around for hundred million. This would entail a large responsibility on governments and civic bodies to provide all communities with safe drinking water as also to plan for industrial and other needs. Water is required every where, without which neither the life nor any development is possible. Public water supply is among the most important needs which mainly caters to the following uses :

(i) Domestic : The use of water for drinking, bathing, watering of lawn and gardens, air conditioning, flushing of toilets, livestocks etc. is designated under domestic use.

(ii) Institutional : Use of water in hospitals, educational institutions, hotel and restaurants, hostels, cinema halls, offices, factories, railway stations etc. is considered as the institutional need of water.

(iii) Public purposes : Water utilized in street sprinkling, watering of public parks, display fountains and cascades, swimming pools and flushing of sewers is considered in the category.

(iv) Fire fighting : Water used in fighting fire menance in both the rural and urban areas is usually considered as the accidental need of the society.

(v) Industrial and Commercial : Apart from above, a substantial quantity of water is lost due to waste through public hydrants and among users.

Quantity of Water Needed

There are fixed Indian standards for the quantum of water to be supplied separately for the aforesaid basic needs, which are as follow ;

Domestic and Non-domestic Need of Water

The following parameters have been recommended depending on the habits and standard of living of the people, provision of flushing systems in residential building and above all the feasibility of supply.

(i) For communities with population upto 10,000	70 to 100 litres/capita
(ii) For communities with population 10,000-50,000	100 to 125 litres/capita
(iii) For communities with population above 50,000	125 to 200 litres/capita

Institutional Need of Water

Requirement of water earmarked for different institutions as enumerated below, will preferably be in addition to the requirements described earlier

Fire Fighting Need of Water

Requirement of water for fire fighting is recommended generally for communities with a population above 50,000. The provision, for this purpose, in kilolitres/day is computed from the formula $100 \sqrt{p}$ where, p is the population in thousand. High rise buildings should have separate fire storage.

TABLE 6.1

Quantity of Water Needed by the Institutions in India

Institutions	*Litres/capita/day*
(i) Hospitals (including laundry)	
(a) No. of beds below 100	340
(b) No. of beds above 100	455
(ii) Hotels	1180
(iii) Hostels	180
(iv) Nurses homes and Medical quarters	135
(v) Boarding Schools/Colleges	135
(vi) Restaurants	70
(vii) Air and sea ports	70
(viii) Junction Stations and primary intermediate Stations	45
(ix) Terminal Stations	45
(x) Day Schools/Colleges	45
(xi) Offices	45
(xii) Factories	45
(xiii) Cinemas/concert halls and Theatres	15

Industrial Need of Water

The water needs for different industries vary widely. Proper assessment will therefore have to be made keeping in view the possible future expansions. As for small industries, other than factories, located insde a town, the per capita supply already enumerated earlier is expected to be sufficient.

TABLE 6.2

Quantity of Water Needed by Various Industries

Industry	*Unit of production*	*Kilolitres per unit*
Automobile	Vehicle	40
Distillery	Alcohol (in kilolitre)	122-170
Fertilizer	Ton	80-200
Leather	100 kg	4
Paper	Ton	200-400
Paper (Special quality)	Ton	400-1000
Straw Board	Ton	75-100
Petroleum Refinery	Ton	1 5-2
Steel	Ton	200-250
Sugar	Ton (can Crushed)	1-2
Textile	100 kg (goods)	8-14

Sources of Water

Water is widely distributed in nature and occurs in number of forms *e.g.*, solid, liquid and vapour. There are two major sources of water *viz.*, surface and ground. Brief account for each of them is given below :

Surface Water

That part of rain water, which flows over the ground to feed rivers, brookes, streams, lakes, ponds etc. is known as run-off or flow-off water.

Rivers and streams are the important surface water resources, through which public water supplies are managed. Quality of surface water depends upon the season of the year, nature of soil and the habitation in the catchment area. But in general, surface water is impure and becomes contaminated where the river flow through thickly populated and industrialised areas. Lakes, natural and artificial, also show similar qualities as in rivers and streams but they, being quiescent, show less pollution both physically and bacteriologically.

Groundwater

Gravity and artesian springs, deep and shallow wells, and infiltration galleries under sandy river beds with a good depth of sand, are the common sources of groundwater. In geological term, the same has been defined as the water that reaches the zone of saturation. It is generally pure when drawn from greater depths as in the case of deep tube wells except that the water may be hard. Public water supplies from groundwater sources are economical as well as less complicated.

Since, the groundwater is found at variable depths, hence, exhibits variable properties. Some of the important terminologies about various types of groundwaters are explained below. Water, which has never been a part of the hydrosphere, is known as Juvenile water. Similarly, a water that has recently been involved in atmospheric circulation is known as meteoric water. The thermal spring water is the transformed form of juvenile into the meteoric water. Besides, there are waters which have been out of contant with atmosphere for millions of years and are known as connate water. The water that has been in association with rocks during their metamorphism is termed as metamorphic water. The water that is derived from magma is known as magmatic water. The magamatic water found at shallow depth is known as volcanic water and that at considerable depth as plutonic water. Many terms are frequently used in connection with the development of groundwater resource management and therefore, their explanations are presented briefly here as follows :

The locus of levels to which water rises in free communication with soil voids is termed as water table. The water table remains constant so long as the withdraw and recharge of the system is maintained in equilibrium. The water bearing strata, capable of yielding sufficiently large quantity of water, so as to be comfortably utilized by the community is termed as aquifer. An aquifer held under pressure by an overlying impervious stratum is known as confined aquifer and as such the water is withdrawn with the help of pump. The aquifer subjected to hydrostatic pressure (as actually happens in case of spring), creates a condition, widely known as artesian, where the flow of water is spontaneously maintained without any pumping device. Water used for the drinking purposes is known as potable water, which must be sparkingly clear, odourless, neither very hard nor too soft and be free from bacteria. Saltish water often fails in quenching the thrust, hence must satisfy all the physico-chemical and bacteriological quantity criteria for human consumption. Waters, on the other hand, used to feed boilers, for washing, and many other non-domestic and industrial purposes are termed as non-potable water.

The fundamental difference between potable and non-potable water lies in the degree of treatment given to them for purification. The waters which are ideally balanced in chemical composition and free from microbial contamination are termed as wholesome water. The suitable treatments needed to make the water wholesome have been thoroughly discussed in subsequent page.

General Impurities Present in Water

Both inorganic and organic impurities are present in the soluble, insoluble and colloidal forms in the water with their corresponding effects, are shown below :

(i) Suspended impurities

(a) Bacteria and Virsus : Some of them cause diseases.

(b) Algae, protoza and fungi : cause odour, turbidity, coluor and cloggihg of pipes.

(ii) Dissolved impurities

(a) Salts

Calcium : Bicarbonate and carbonate-Alkalinity and Hardness

Magnesium : Sulphate-Hardness

Chloride : Hardness, Corrosive to boilers

Sodium : Bicarbonate and Carbonate-Alkalinity

Sulphate-Forming in boilers

Chloride Brakish taste

Iron : Oxide-Bad taste, red water, discolouration of clothes corrosive of metals, hardness

Manganese : Black or brown water, discolours of clothes, helps organic growth

Lead : Cumulative poison, causing lead poisoning.

(iii) Gases

(i) Oxygen : Corrosive to metals.

(ii) Carbon dioxide : Acidity, corrosion.

(iii) Hydrogen sulphide : Odour like rotton egg, acidity.

(iv) Nitrogen : Albuminion ammonia-Vegetable pollution, Fee ammonia sewage pollution.

TABLE 6.3

Physico-chemical Limits for Drinking Water

Parameter	*Acceptance*	*Cause for rejection*
Turbidity (J.T.U. scale)	2.5	10
Taste and odour	Objectionable	Objectionable
Total dissolved solids (mg/l)	500	1500
Chloride (as Cl) mg/l	200	1000

Flouride (as F) mg/l	1.0	1.5
Calcium (as Ca) mg/l	75	200
Iron (as Fe) mg/l	0.1	0.1
Copper (as Cu) mg/l	0.05	1.5
Phenolic compoudns (as phenol) mg/l	0.001	0.002
Toxic Materials		
Arsenic (as As) mg/l	0.05	0.05
Chromium (as hexavalent Cr) mg/l	0.05	0.05
Lead (as Pb) mg/l	0.1	0.1
Mercury (total as Hg) mg/l	0.001	0.001
Gross Alpha-activity (pcl)*	3	3
Colour (platinum cobalt scale)	5.0	25
pH	7.0 to 8.5	6.5 to 9.2
Total hardness (mg/l) as $CaCO_3$	200	600
Sulphate (as SO_4) mg/l	200	400
Nitrate (as NO_3) mg/l	45	45
Magnesium (as Mg) mg/l	30	150
Manganese (as Mg) mg/	0.05	0.5
Zinc (as Zn) mg/l	5.0	15.0
Anionic detergents (as MBAS) mg/l		
Toxic Material	0.2	1.0
Cadmium (as Cd) mg/l	0.01	0.01
Cyanides (as CN) mg/l	0.05	0.05
Selenium (as Se) mg/l	0.01	0.01
Polynuclear aromatic hydrocarbons		
Radio activity	0.2/ug/l	0.2/ug/l
Gross Beta activity (pcl)*	30	30

pcl* = pico curie, equal to 10^{-12} cl

Whereas 1 cl = 3.7×10^{19} atoms disintegration/sec (dps)

= 2.2×10^{11} dpm.

Conventional Treatment Methods for Public Water Supply

Conventional methods for water supply scheme mainly achieve the following purposes by involving a number of step. Some of them are given below :

(a) Removal of colour, odour and taste

(b) Removal of excessive hardness

(c) Removal of iron and manganese

(d) Maintaining flouride level

(e) Demineralisation if needed

(f) Chlorination.

The steps taken in raw water treatment is presented below :
Raw water → PC → A → Fl → POC

Where PC : Pre-chlorination

A : Aeration

F : Flocculation

Fl : Filtration

POC : Post-chlorination.

The sources and the quality of raw water determine the type of treatments to be undertaken. For example, if the water is drawn from deep tubewells, coagulation and filtration are not necessary. Only softening may be necessary where the water is hard, or aeration where the water contains excessive iron taste or odour. For surface water, flocculation and disinfection are essential whereas prechlorination and aeration may be necessary in specific cases, and especially during rain season.

Aeration

The principle behind this method is to mix air with water to increase the oxygen content and remove carbon dioxide. Aeration is done to remove iron from groundwater and taste and odour from surface water. It is mostly done in the initial stage of water treatment. It is a simple process and is accomplished by the following methods :

(i) Cascades-Water is made to fall in a thin sheet down a series of steps like a fountain.

(ii) Spray Nozzles-Water is sprayed under pressure to open air through nozzles.

(iii) Waterfall or Multiples Tray Aerators-Water is discharged through a riser pipe and allowed to fall on a series of trays from which it falls either through small openings or over the edges.

(iv) Mechanical Aerators-These are most applied in waste water treatment, where a large amount of organic matters are to be oxidised, within a short period of time. This method is suitable also for the area where land is a limiting factor.

Coagulation and Flocculation

Finely divided suspended matters, both organic and inorganic which cause colour and turbidit . do not settle easily through sedimentation in a short period, the object of coagulation is to convert these into easily settleable size, by reducing the zeta potential forces. The latter are repulsive forces which on reduction, bind together the colloidal particles. Thus, the aggregated molecules of suspended colloidal particles become large enough in size and settle down easily. When the coagulants are added to water, large size flocks are formed on rapid agitation. These flocks absorb colour and during their downward journey entrap suspended matter and bacteria. This process is known as flocculation.

The most common coagulant is aluminium sulphate or $Al_2(SO_4)_3$, $18H_2O$ and the next is ferrous sulphate, $FeSO_4$, $7H_2O$.

Alum is preferred because it requires lesser skill and control. The reactions that take place are as such :

$$Al_2(SO_4)_3+18\ H_2O+3\ Ca(HCO_3)_2=2Al(OH)_3$$
$$+3CaSO_4+18H_2O+6CO_2$$

In above reaction $Ca(HCO_3)_2$ represents the natural alkalinity of waters. Some waters which do not have sufficient alkalinity need the addition of lime. CaO (lime) react with water to form calcium hydroxide, $Ca(OH)_2$ which further reacts with alum as follows :

$$Al_2(SO_4)_3+18H_2O+3Ca(OH)_2=3Al(OH)_3$$
$$+3CaSO_4+18H_2O$$

In both case the flock aluminium hydroxide is formed as coagulation depends on the pH of water. Different coagulants may be suitable for different pH range. Alum is most commonly used in public water supply scheme as coagulant. The use of alum in the form of solid in water treatment process is not desirable at all, because as such, the natural buffering capacity of aluminised water becomes locally upset, which in turn, produces excessively high turbidity.

TABLE 6.4

Commonly use Coagulants to Raw Water Treatment

Common Coagulants	*pH range*
Alum	6.5 to 8.0
Ferrous Sulphate	8.5 to 11.0
Ferric Sulphate	5.0 to 11.0
Ferric Chloride	5.0 to 11.0
Sodium Aluminate	6.0 to 8.5

Thus, for efficient coagulation and flock formation, optimum pH, adequate does, rapid agaitation and gentle transfer of flocculated water to selling tank is necessary.

Quality Creteria for Alum and Lime as Coagulants

Good quality of alum must contain alumina (Al_2O_3) in the range of 16 to 17%. Besides there must not be more than 0.7% of insoluble matter and 0.3% iron. The pH value of 10% aqueous solution of alum should vary in the range of 2.7

to 3 2 (IS 299-1975) otherwise natural alkalinity of the water to be coagulated would be severely damaged resulting in very high consumption of lime. Relaxation has however, been granted by Indian Standard Institution, Government of India. by reducing the limit of alumina to 15% for the use as coagulant especially in public water supply purification process (IS 299-1980). In case of coal washery wastes the use of alum as coagulant does not achieve the desirable degree of coagulaltion. For effective coagulation of the suspended coal particulates, recently developed poly-electrolyte in very little dose of 0.3 to 0.7 ppm in combination with alum and lime is of great success. The poly-electrolyte as such whorks as coagulantaid.

Lime needs to be of equally good quality otherwise effective coagulation is not possible. Good quality of lime must have the minimum 90% content of available lime as CaO (IS-1540-1978). Too high content of lime as CaO in lime sample is to good to be used in river water treatment rather is wasteful in the case where natural alkalinity of raw water is more than 80 ppm. Experiments have conclusively proved that lime within 78.6% as CaO is good enough to be used in water purification, especially where natural alkalinity of water ranges from 65 to 110 ppm. Further the content of CaO in lime is shown to be considerably reduced on ageing, especially when the same is stored in the open air, owing to the conversion of CaO into $CaCO_3$. Approximatelly, 50% of its total CaO content is lost during one and half month storage in the haphazard manner. Insoluble matter present in the lime as silca is though permitted upto 1.0 per cent (IS 1540-11-1971). but practically of no harm, even if it exceeds a number of times. This is rather helpful in reason where the hardness of water is found usually in the range of 100 to 125 ppm. At the same time, it should always be in mind that the insoluble matters, other than silica, are not desirable, because as such, the treated water becomes even more turbid than the raw water itself. Therefore, special care is needed in the identification of the types of insoluble matters present as the impurities in both alum as well as in the lime.

For quantitative analysis of alum and lime, well equipped laboratory is needed. Special care in weighing the samples and making volumetric solutions is of vital importance in achieving the accuracy in the result. The details of the procedures for the estimation of alum and lime are described in the IS 299-1975 and IS-1514-1059 respectively.

Filtration

Filtration aids in the removal of taste, odour. colour, turbidity, iron and manganese and to a large extent of bacterial load. Three types of filters are usually used in water treatment plants. These are:

(i) Rapid stand filter,

(ii) Slow sand filter and

(iii) Pressure filter. Relative merits and demerits for each of them are discussed as follows;

Rapid Sand Filter

Essential characteristic of rapid sand filters are high rate of filtration and consequently lesser area requirement and washing the filters through reverse flow of filter water through the filter beds. A rapid sand filter consist of a bed. supported by gravels of different sizes in layers. Filter sand for a rapid sand filter should satisfy the following norms:

(a) Sand shall be hard, resistant and free from dirt of every description

(b) Effective size should range between 0.45 to 0.70 mm

(c) Unformity coefficient must not be more than 1.7 and less than 1.3

(d) Ignition loss not be exceed 0.7 per cent by weight

(e) Not more than 5 per cent of the sand be soluble in HCl.

(f) Silca content should not be less than 99.8 per cent

(g) Specific gravity to be in the range of 2.55 to 2.65,

The depth of filter sand layer should be between 60 to 75 cm. The standing depth of water above filter sand must be between 1 to 2 metres. The free board above water level should be at least 50 cm. Besides these, the operation and maintenance of rapid sand filter need the strict implementation of the following knowhow :

The standard rate of filteration through a rapid sand filter is usually 80 to 100 lpm/m² (litres per minute/square meter). High rate can be achieved by proper conditioning of water before filtration. For plants of 1000 md (million litres a day) capacity or more, 100 m² area is needed to be provided in two units of 50 m² each. Besides the under-drain consists of a grid of manifolds, laterals, headers covering the entire filter bottom to collect filtered water and also to distribute wash water. These are perforated and pipes of cost iron, asbestos cement, concrete or other material.

For perforated pipe under-drain system, minimum gravel size should be 2 mm and maximum size 25 mm. Depth of gravel should be 50 cm. Filter gravel shall be spherical, hard, clean and uniform in size. Gravel is placed between the sand and the under-drainage system, the lower size being at the top and higher at the bottom above the under-drains.

It is essential to wash the filter beds at regular intervals to prevent clogging. Washing is done by wash water under pressure, sent through the underdrainage system into the filter beds. Pressure should be such as not to disturb the sand beds. Normel rates of application is 600 lpm/m², equivalent to a rise of 60 cm/min for a period of 10 minutes. Air wash system is used to precede water wash of the beds. The combined systems of washing reduces the use of wash water, therefore, very useful in case of large filter units. Air is pushed through the unerdrains for about 3.5 minutes to agitate the sand beds before introducing wash water.

The disposal of the waste generated during water treatment process is itself a major task for the sanitary engineers and chemists. The solid wastes, produced in the water treatment plants mainly consist of following composition :

(a) Sludge from flocculation, residual chemical coagulants plankton etc.

(b) Wastes from backwashing of filter beds.

(c) Wastes generated during ion-exchange water softening process.

(ii) Slow Sound Filter

The erection of this type of treatment plant involves initially very high cost and need a large area than that of rapid sand filter process. The effective size of sand ranges from 0.2 to 0.3 mm. This process is very efficient in removing bacteria, colour and odour, even without the use of any chemical *e.g.* alum, lime and bleaching powder. However, this process is not at all suitable for the water which has turbidity value greater than 50 ppm and involves a large number of people. This process in fact is meant for the rural area where electricity is a problem and the availability of land is not a limiting factor.

The comparative merits and demerits for the rapid and slow sand filters have, however, been shown in the following table for assessing their feasible at a glance.

TABLE 6.5

Comparison of the Rapid and Slow Sand Filters

Parameter	*Slow Sand Filter*	*Rapid Sand Filter*
Compactness	Requires large area	Small area requirement gives compactness.
Chemical treatment	Not generally required	Essential
Filter media	Fine sand, effective size 0.2 to 0.3 mm	Coarser sand effective size 0.45 to 0.75 mm.

Filtration rate	2 to 3 lpm/m²	80 to 100 lpm/m² may be increased if necessary
Flexibility	Not suitable for variation of demand	Suitable
Period of run	1 to 3 months	1 to 3 days depending upon the quality of raw water
Supervision	Skilled supervision not necessary	Skilled supervision essential
Economy	Initial cost very high	Economical
Suitability	Not suitable for water with turbidity above 50 ppm	Suitable for turbid and coloured water
Efficiency	Very efficient in removing bacteria, Chlorination not necessary	Bacteria removal 90-95% Colour and taste are also emoved. Chlorination essential

(iii) Pressure Filter

Pressure filter is a type of rapid sand filter in a closed container through which water is passed under pressure. It is not considered very reliable in removing bacteria. Therefore, it is better used to soften and remove iron from groundwater. It has a wide use in clarifying softned water in industrial plants. Nevertheless pressure filter is used in small water works in our country. A battery of units are installed, as per requirement, all receiving raw or chemically treated water and discharging into a common header pipe or pipes. Vertical pressure filters are generally 16-96 inch in diameter, while horizontal units are obtainable in 7 to 8 ft. diameter length varies from 8 to 25 ft.

Disinfection

Treatment of water by coagulation, sedimentation and filteration renders water chemically and aesthetically acceptable,

Although a large part of the bacterial load is reduced during such treatment even then the water does not become absolutely free of them. Water, therefore, requires disinfection which implies the destruction of pathogen, *i.e.*, (disease producing) bacteria.

Disinfection does not completely destroy all micro-organisms, which can be accomplished only by sterilisation. Complete sterilisation however, is not easily obtained, nor it is necessary, for making water potable. The commonly used chemicals and disinfectants are bleaching powder, liquid chlorine iodine, etc. Thelater is available in the form of tablets, known as tetraglycine hydro-periodide, galabaline, etc., iodine is used in disinfecting the water in such places, *e.g.*, in the field of war, where no organised water supply facility exists.

(i) Chlorination

Chlorine not only removes bacteria but also eliminates taste, colour, and odour controls algae, protozoa and fungi, improves coagulation and oxidises iron and manganese. It is available as calcium hypochlorite in form of bleaching powder containing 33-35 per cent available chlorine (IS : 1065.1971) and in the form of HTH (High test hypochlorite) powder which is stable carries 65-75 per cent of available chlorine. These are used in small water works. For medium and large water works, chlorine gas is used because it is more economical and effective. It is compressed into liquid form and supplied in steel cylinders and is injected in gaseous state through chlorinators.

Chlorine reacts with water to form hypochlorous acid (HOCl) and hydrochloric acid (HCl). The reaction is represented as :

$$H_2O + Cl_2 = HOCl + H^+ + Cl^{--}$$

Hypochlorous acid being unstable dissociates into hydrogen ions (H^+) and hypochlorite ions (OCl) pt.

$$HOCl = H^+ + OCl$$

Both the above equations are reversible, the undissociated HOCl is 80 to 100 times more potent as a disinfectant than the

OCl ion. The amount of dissociation is directly dependent upon the pH value of water. The disinfecting action is rapid when the pH value is about 7 or slightly higher. Beyond, it is reduc d gradually because more and more HOCl dissociates into OCl ions. Equal amount of HOCl and OCl is present at pH 7.5.

Chlorine present in water as hypochlorous acid, hypochlorite ions and molecular chlorine is defined as free available chlorine. Normal chlorine dose varies from trace to 1 ppm depending on the pollution load. The dose is determined with precision because disinfection will not be complete with insufficient dose, whereas a higher dose will form bad taste and odour, necessitating dechlorination. An effective does of chlorine will leave a residual chlorine charge of 0.1 to 0.2 ppm which indicates that the water has been properly disinfected.

Chlorine, when added to water as free chlorine, oxidises some of the organic and inorganic compounds. While reacting with ammonia and organic amines, it forms chloramines. Chlorine present in water in chemical combination with amonia or nitrogenous compounds is known as combined available chlorine.

Chloramines are also oxidising agents but are less potent than hypochlorous acid. The following reactions mark the formation of the three chloramines :

$$NH_3 + HOCl = H_2O + NH_2Cl \text{ (monochloramine)}$$

$$NH_3Cl + HOCl = H_2O + NHCl_2 \text{ (Dichloramine)}$$

$$NHCl_2 + HOCl = H_2O + NCl_3 \text{ (Trichloramine)}$$

Monochloramine predominates at pH over 7 5 dichloramine at pH between 5 to 6.5 and odorours trichloramine (nitrogen trichloride) is produced at pH below 4.4. It is, however, volatile and disappears within a few hours retention period, but is carcinogenic in nature. The chlorination of water to the extent that all the ammonia be converted into trichloramine (NCl_3) or oxidised to free nitrogen is termed as break-point chlorination.

This is estimated by knowing the sudden fall in residual chlorine (RC) value inspite of continued chlorine addition to the same volume of water under test. Similarly, the difference between the amount of chlorine applied to the constant volume of water (say of one litre) and the amount of free, combined or total available residual chlorine remaining at the end of the desired contact period is known as chlorine demand of water. The latter determines the quality of raw water and thus, bears the positive relationship with the amount of oxidisable organic pollution load present in water. Raw water with a chlorine demand upto a value of 1.1 ppm is considered to be fit to be used as source of public water supply.

To ensure the wholesomeness of the drinking water, accurate estimation of free, residual, combined residual and total residual chlorine is essential.

The instrument used for chlorine estimation on spot is of pocket size known as chloroscope. Its manufacture has recently been commercialized in India and it is easily available from any scientific instrument dealer.

(ii) Methods for estimation of Available Chlorine in Water

Put 0.5 ml of 0.2 per cent orthotoluidine solution + 10 ml water sample + 0.5 ml of 0.5 per cent solution of sodium arsenate ($NaASO_3$).

(a) Read immediately and discard-A

(b) Put 0.5 ml $NaASO_3$ + 10 ml water sample + 0.5 ml orthotoluidine solution.

Read immediately B_1.

Read after 5 minutes-B_2.

(c) 0.5 ml orthotoluidine + 10 ml water sample, Read immediately C

Total residual chlorine = C-B_t,

Free residual chlorine = A-B_1,

Combined residual = (C-B_2) - (A-B_1)

At 20-25°C and pH 6.8, 0.2 ppm of free residual chlorine (FRC) for 10 minutes contact period is considered as the optimum dose for killing most of the bacteria or rendering them ineffective. The optimum FRC concentration (0.2 ppm) at the furthest points of the distribution network can only be maintained, provided, the following three points are satisfied.

(a) FRC should range from 0.5 to 1.2 ppm at the source.

(b) The distance between the source and the furthest distribution must be preferably within 10 kms, measured along the delivery pipes. For longer distribution system, re-chlorination will have to be done at suitable point to maintain requisite FRC.

(c) The minimum pressure 1.4 kg/cm^2 must be maintained throughout the distribution system, otherwise back flows of polluted water may occur.

(iii) Estimation of Available Chlorine in Bleaching Powder

Prepare 1 per cent aqueous solution of bleaching powder. Out of the total prepared solution, 10 ml is taken out and to which 5 ml acetic alongwith 2 g solid potassiu n iodine are added. It is then titrated against N/40 sodium thiosulphate ($N_3aS_4O_3$) solution, using starch as indicator, till the end point comes. The calculation for the available chlorine is done as indicated below :

$$\% \text{ Available chlorine} = \text{ml of N/40 } Na_2S_4O_3 \times 0.89.$$

(iv) Other Methods of Disinfection

Boiling

This is the easiest method to destroy all kinds of bacteria, alage etc. It needs vigorous boiling for 15 to 20 minutes. It is applicable for individual use because large volumes of water cannot be treated by this method. During epidemic,.water must be consumed after boiling, even if the same is chlorinated, to become doubly sure about its sterlity.

Ultraviolet Irradiation

This is the most modern of sterlisation and consists of passing ultra violet rays from mercury vapour arc lamps. The astonishing results of this method is shown here.

TABLE 6.6

Killing time for Pathogeninc Bacteria

Pathogens	*Time (sec)*
Cholera	10 – 15
Dysentry	13 – 20
Tetanus	20 – 60
Staphylococeus	5 – 10

Smoe Special Treatments for Ground Water

As has earlier been said, groundwaters are usually more hard as compared to the surface waters. Besides, the groundwater in some regions contains large proportion of dissolved elements *viz.* Iron, manganese, flouride, chloride, iodine, etc., which necessitate removal.

The groundwater in the extreme southern parts along the plateau regions across the border of Orissa, Madhya Pradesh and Bihar is acidic and as such need to be treated.

Hardness

There are two types of hardness, temporary and permanent. Temporary hardness is caused due to the presence of carbonate and bi-carbonate of calcium, magnesium and other bivalent metals. It can be removed by boiling. This is known as carbonate hardness. Permanent hardness is caused due to the presence of sulphates and chlorides of hardness constituting cations. This is also known as non-carbonate hardness and cannot be removed by easy means like boiling. These salts are very corrosive and deposit scales in steam boilers and in distribution system.

Waters having hardness values of less than 45 ppm is known to be very soft water and with more than 230 ppm as excessively hard water. The consumption of very soft water, especially in old age is desirable, because the toxic elements are removed from the body. The use of very hard water causes the loss of the tenderness of vegetables and meat cooked. Possible inflamation of intestines have been reported to be lessened or removed by consuming moderately soft wate.r

TABLE 6.7

Classification of Waters on the basis of their Hardness Value

ppm	*Degree of Hardness*
15	Extremely soft
30	Very soft
45	Soft
90	Moderately soft
110	Moderately hard
130	Hard
170	Very hard
230	Excessively hard
250 and above	Too hard for use

Removal of Temporary Hardness

(i) *Boiling* : Soluble Ca and Mg bio-carbonates are converted into their insoluble carbonates. The mechanism of the reactions is as shown below :

$$Ca(HCO_3)_2 = CaCO_3\downarrow + H_2O + CO_2$$

$$Mg(HCO_3)_2 = MgCO_3\downarrow + H_2O + CO_2$$

(ii) *Clark's Process* : Water is softened by the addition of necessary amount of lime in the form of lime water of milk of lime.

$Ca(HCO_3)_2$ is precipitated as $CaCO_3$. In case of $Mg(HCO_3)_2$ lime is added in double the quantity to form sparingly soluble $Mg(OH)_2$.

$$Ca(HCO_3)_2 + Ca(OH)_2 = 2CaCO_3\downarrow + 2H_2O$$

$$Mg(HCO_3)_2 + 2Ca(OH)_2 = 2CaCO_3\downarrow + Mg(OH)_2 + 2H_2O$$

Lime softened water is not stable and causes deposits of calcium carbonate in pipes and filter sands. Addition of alum stabilises it.

Removal of Permanent Hardness

(i) *Soda-Lime Process* - Permanently hard water can be softened by the addition of sodium carbonate (washing soda). It converts the soluble sulphates and chlorides of Ca and Mg into insoluble carbonates of Ca and Mg Water cannot be softened by this process below the hardness value of 21 ppm.

$$CaCl_2 + Na_2CO_3 = CaCO_3\downarrow + 2NaCl$$

$$MgSO_4 + Na_2CO_3 = MgCO_3\downarrow + Na_2SO_4$$

(ii) *Ion-Exchange Method*—Water of zero hardness value can easily be produced by this process Zeolites are used as the softener which are complex compounds of sodium, aluminium and silica. These are either artificially prepared or may be obtained from natural deposits and refined. They can easily exchange bases. The principle behind the method is that when hard water (both temporary and permanent) is filtered through zeolite, the calcium and magnesium cations are removed in exchange of sodium. The filtrate being free from Ca and Mg salts is soft. When the sodium of the zeolite is exhausted, it is regenerated by applying salt solution whereby an exchange is again affected. Softeners of the type are usually constructed like pressure filters.

System of Distribution

Distribution of water is done either by intermittent or continuous system In the intermittent system, wastage of water may be controlled but the main disadvantage is that during non supply hours, water in the pipeline rushes towards the lower reaches, thus creating partial vacuum in pipes, resulting in the sucking of impurities and gases in the pipeline. This may cause health hazards. Also during times of fire, intermittent supply system proved to be a great handicap.

Continuous supply, though preferable in everyway, but cannot be functional in every town, unless supply to every household is metered to avoid wastage. Grid system of distribution mains where all the pipelines are interconnected is always preferred. It eliminates dead ends.

Zoning

Zoning to the distribution is done for equitable distribution of water throughout the area. It mainly depends on the topography, density of population and type of locality.

Service Reservoirs

It is of importance for regulation for pressure in the distribution system. It also regulates the supply in case of failure of electricity, pumps or any other such defects. The elevation of the reservoir should be designed to maintain minimum residual pressure in the distribution mains. In towns of undulating topography as well as in big towns, several service reservoirs are constructed at suitable places in different zones.

Pipe Size and Residual Pressure

For towns having population up to 50,0.0, minimum pipe size shall be .00 mm, whereas for population above 50,000 it should be 150 mm.

Minimum residual pressure at ferrule points in a distribution system shall be :

For single storeyed building	7 m
For two storeyed building	12 m
For three storeyed building	17 m

For higher buildings, boosting of water should be done by booster pumps.

Laying and Maintenance of Mains

For roads wider than 25 meters, distribution pipe shall be laid on both sides provideded with suitable valves. Sluice valves should be provided at regular intervals and at junctions to isolate the portions for repair, when necessary. Air valves are provided to release air automatically when pipelines are filled. They are provided at the peak points in the pipeline gradient to release entrapped air. Absence of air valves may result in deficiency in flow or even bursting of pipes. Scour valves are provided at the valley points. They are opened at regular intervals to cleanse the pipeline while water rushes out to empty the pipe line for repair or replacement. The use of hemp yarn while caulking with lead of cast iron pipe sometimes infects the water. The usual practice is to use bituminous coating as pipe lining materials is identified to increase the concentration of Poly nuclear aromatic hydrocarbon (PAH . Now-a-days, cement lining inside and zinc coating outside, of the cast iron pipe is gaining importance for smooth surface and longer working life. The latter is high accumlating in nature and known to be teratogenic (a substance which induces a change in the genes). The use of Poly Vinyl chloride (PVC) pipe in public water supply is also not good, because it contributes lead by leaching to the water if not properly controlled. PVC is not good for hot water supply.

Water supply to any individual building is made through house service connection. It is guided by the bye-laws of the concerned municipality.

TABLE 6 8

Langelier Co-efficients

TDS ppm	*A*	*Calcium Hardness or 2/3rd total harduess (as $CaCO_3$) ppm*	*C*	*Total alkalinity (as $Ca\,O_3$) ppm*	*D*
1	2	3	4	5	6
50 - 300	0.1	10–11	0.6	10–11	1.0
440–1000	0.2	12–13	0.7	12–13	1.1
		14–17	0.8	14–17	1.2
		18–22	0.9	18–22	1.3
		23–27	1.0	23–27	1.4
		28—34	1.1	28—35	1.5

(*Contd.*)

1	2	3	4	5	6
Temperature (°C)	B				
0–1	2.6	35–43	0.2	36–44	1.6
2–6	2.5	44–55	1.3	44–55	1.7
7–9	2.4	56–69	1.4	56–69	1.8
10–13	2.3	70–87	1.5	70–88	1.9
14–17	2.2	88–110	1.6	89–110	2.0
18–21	2.1	111–138	1.7	111–139	2.1
22–27	2.0	139–174	1.8	140–176	2.2
28–31	1.9	175–220	1.9	177–220	2.3
32–37	1.8	230–270	2.0	230–270	2.4
38–43	1.7	280–340	2.1	280–350	2.5
44–50	1.6	350–430	2.2	360–440	2.6
51–56	1.5	440–550	2.3	450–550	2.7
57–63	1.4	560–690	2.4	660–690	2.8
64–71	1.3	700–870	2.5	700–880	2.9
72–81	1.2	880–1000	2.6	890–1000	3.0

Wastage of Water

Wastage may occur in treatment plant, from leakages in service reservoir and pipelines, careless use by consumers through continuous flow in public hydrants, due to theft of cocks and so on Levels of wastage upto 10% may be considered low but above it, corrective measures are needed from the point of view of economy.

Protection against Pollution Near Sewers and Drains

Water main should be at leats 3 m away from the existing or proposed sewer or drain. In case where it has to cross huse or sanitary sewer of storm drain the bottom of the main should be at least 0.5 m above the top of the drain or sewer.

Incrustation and Corrosion of Delivery Pipes

Incrustation and corrosion are the major enemies of distribution pipes and valves, coausing tremendous national loss in term of money and loss of out put. Incrustation significantly reduces carrying capacity of the water delivery pipes. Considerable amount of energy is lost due to incrustation in boilers. Similarly the aesthetic value of drinking water is tremendously lost due to incrustation and corrosion of cost iron pipes. The life of shallow hand pump pipes is drastically reduced from average length of 30 years to less than 5 years in regions where water is either acidic or contains exceedingly high proportion of soluble iron. Keeping in view the tremendous loss, caused mainly by incrustation of corrosion, an urgent need to envisage a proper check and balance method has been felt throughout the world. A laboratory method to ascertain the nature of water, along with easily applicable methods for the control are briefly presented here.

Laboratory Method to ascertain the Corrosive and incrustating Nature of Water

Langelier Index : $LI = pH - pHs$ (1)

Where, pH is actual pH of water and pHs = Saturated pH produced by adding calcium carbonate.

The Langelier inded (LI) with zero value indicates that water is neither corrosive nor incrusting in nature *i.e.* the water is chemically balanced. A positive value indicates the tendency of water toward incrusting. A negative value indicatesa tendency to eat up the metals *i.e.* water under examination is corrosive in nature. The value of the saturation pH in equation (1) may be determined as such

$$pHs = 9.3 + (A + B) - (C + D) \quad \text{.. (2)}$$

The co-efficients A, B, C and D indicate the following factors

(i) Factor A-TDS

(ii) Factor B Water temperature

(iii) Factor C-Calcium carbonate in ppm

(iv) Factor D-Total range alkalinity in ppm as calcium carbonate.

Prvention of Incrustation

Incrustation cannot be prevented entirely, but it can be delayed and kept in check by keeping the draw-down as small as possible, as follows :

(i) In order to reduce the head loss to a minimum the well must be developed properly so that aquifer losses are kept at its minimum

(ii) A screen with large open area and fully penetrating the aquifer should be installed. As such, the high precipitation of iron salts and carbonate will be reduced

(iii) The pumping rate should be reduced by increasing the pumping period. The required quantity of water may be obtained from several wells rather than pumping from a few large wells at excessively higher rates

(iv) The screens should be cleaned periodically, at least once in a year, even if the discharge has not fallen-off.

In case of already incrustated wells where treatment is essential, the same may be achieved by using acid, chlorine or other effective agents, depending upon the nature of the incrustated materials. Hydrochloric and sulphuric acids are effective in removing carbonates. Sodium hexa meta phosphate is effective in dispersing iron and manganese oxides. Chlorine is effective in removing bacterial growth and slimes.

Corrosion

The negative Langelier index results in corrosion which in turn, reduces the strength of the pipe, increases the screen size, causing sand pumping and eventual failure of the tubewell. The important reasons for corrosion are galvanic action corrosive water, high velocity and unsaturated waters.

(i) ***Electrolytic or Galvanic Action***– It is the most important factor leading to corrosion. The principle behind electrolytic corrosion is that when two or more dissimilar metals are dipped into solution (*viz.* in water) a solution potential is set up which causes the electrons to move from cathode (electrode) to anode (electrode). Thus the anodic portion is corroded and deposition takes place on the cathode. The elements lying above the hydrogen in electrochemical series, act as anode and those below as cathode. The positions of the elements in electrochemical series determines the rate of corrosion, depending upon their mutual positions. Generall, the greater the difference in their position, more intence corrosion is produced. For example, if iron and brass are coupled together in water, iron acts as anode and brass as cathod. Thereby, iron is progressively being corroded and brass remains protected. Sometimes, corrosion takes place in the same metal between two path of slightly dissimilar composition. This is known as local corrosion. The protection of iron pipe is achieved by galvanizing the same. In doing so, the zinc start working as anode and iron as cathod. Thus, zinc is corroded and iron is protected from corrosion.

(ii) *Corrosive water*– If the water is acidic, it attacks the metals diJectly and causes corrosion. Hydrogen sulphide corrodes steel and copper alloys, if it is present in quantities greater than 1 ppm. If the dissolved carbon dioxide concentration is more than 53 ppm and DO is 2 ppm. make the water corrosive. The electyical conductivity value above 1500 mhos/cm makes water corrosive.

(iii) *High velocity of water* – The high flow of water removes the hydrogen from the cathode, which in turn accentuates the process of corrosion. Very low velocity of water tends to prevent local corrosion but makes it uniform.

(iv) *Unsaturated water* - When water is deprived of the minimum quantity of iron that a water can normaly contnin, It is known as unsaturated water. As such, the water corrodes iron and steel until it becomes saturated. The water with less than 50 ppm of hardness is generally unsaturated and often causes corrosion.

Methods to Control Corrosion

(i) The use of corrosion resistant materials can effectively control corrosion. Such materials include everdur bronze (16% copper, 3% silicon, 1% manganese) stainless steel, polythene, epoxy-bonded fibre glass etc.

(ii) Galvanzing, coating with plastics etc. offer considerable protection from corrosion.

(iii) Reduction in velocity reduces the rate of corrosion. It can be achieved by increasing the percentage of open area of the diameter of the well pipe or decreasing the rate of pumping or the draw down.

(iv) The corrosiveness of water may be affectively reduced by suitably treating such water. The addition of lime to neutralise the acidity of water is worth suggesting.

(v) Cathode protection.

In this method an alternative anode is selected. Thus, the iron pipe may be protected from corrosion by making zinc as anode.

7

Industrial Pollution Abatement : Some Case Studies

Introduction

There are broadly, two types of pollution from industries which need attention as far as their treatment and disposal is concerned-water and air pollution. Treatment technology of domestic-sewage liquid waste is well developed; as the composition of the waste is more; or less constant. Industrial waste waters vary widely in their composition and treatment methods, which have to take into consideration the specific characteristics of the wastes. Many treatment practices have followed the approach of mixing the liquid-sewage waste; with; industrial waste and treating the mixture by biological oxidation, broadly as followed for; sewage; waste. A number of; treatment plants have come up in India for the treatment of petrochemical, pharmaceutical, dairy, photofilm, synthetic-drug and antibiotic wastes. Low-cost methods of treatment, such as lagoons (aerobic and anaerobic), oxidation ditches and aerated logoons have also been tried with varying degree of success has been reported in biological-treatment plants for liquid wastes emanating from food and fruit processing units, dairies, slaughter houses and textile industries. Stabilization ponds, activated or modified-activated sludge processes, aerated lagoons, trickling filters and anaerobic biological-digestion units have been used for treating these wastes. But the fact remains that biological treatment, even though cheap in capital and operating costs, may not be a panacea for industrial waste-water problems. Physio-chemical methods are necessary to remove or recover the chemical ingredients present in liquid effluents discharged; from electroplating, chlor-alkali, pesticide, fertilizer, dyes and pigme-ments, metallurgical, paper-and-pulp and other such process industries.

The re-use of water in processes where the water quality standards are not stringent is worth considering. A considerable quantity of water is presently being se-used in process industries in India but a lot more needs to be done in this area. Tertiary or advanced methods of treatment that are essentially physico-chemical should be developed and practised on a large scale in order to improve the quality of water for re-use or even drinking. This becomes important as fresh-water resources are getting depleted fast and may have to be reserved for drinking alone.

A survey of the air quality in major cities in India such as Bombay, Calcutta, Delhi, Kanpur and Nagpur, for which detailed ambient-quality monitoring has been done over extended periods, indicates that the levels of sulphur dioxide, particle matter and oxides of nitrogen are much higher than those specified by the Environmental Protection Agency (EPA). The pollution level in ambient air can best be reduced by reducing concentrations in effluent gases from process industries.

Pollution Control Aspects of Fertilizer Industries

The world population is increasing at a rapid rate. More chemicals and fertilizers will be required to meet; the increased food-production requirements caused by this increase. More industries, if required to be set up, would narrow the gap between residential areas; and factory locations, more; so in urban areas. The close proximity of residential areas and factories makes the effluent-disposal problem more difficult. This problem becomes much more acute for highly populated areas such as Calcutta, Bombay and Delhi where the emission limits will have; to be made stringent as the population of these cities will be susceptible to the multipronged and cumulative effects of pollution. Science and technology have, therefore, to be kept abreast of these developments and newer methods should be developed to meet the demands of stricter controls that will become necessary in future. The wastes of yesterday could become useful products of today. Newer dimensions will be added to the twin problems of pollution control and, utilization of wastes from gaseous, liquid or solid effluents.

A fertilizer unit of average size, in India, producing nitrogenous fertilizers, would consist of an ammonia plant (for conversion into urea, ammonium sulphate, nitrolime or their combinations). A unit producing phosphatic fertilizers would consist of nitric acid (for nitrophosphate production), phosphoric acid and sulphuric acid plants. A complex fertilizer plant producing balanced fertilizers NP or NPK shall have all the above mentioned processes. The nature of wastes or pollutants normally depends on the product mix that is being handled, in a particular, unit and varies widely from plant to plant, the liquid effluent can no longer be discharged freely into rivers, seas or on the land as was hitherto the practice. Strict regulations are passed by state and central governments, and, enforced by their, pollution boards, to ensure that industrial effluents fall within their prescribed limits. A typical fertilizer unit in India Produces 300 tonnes/day of ammonia, 300 tonnes/day of urea, 1100 tonnes/day of complex fertilizers, and 130 tonnes/day of methanol. About 10,000 m^3/day of liquid effluent has to be suitably disposed. The liquid containing both contaminated as well as uncontaminated components can weigh up to 50,000 tonnes/day. Fertilizer plants in India can be classified into the following five categories :

1. Ammonia and urea plants
2. Ammonia, nitric acid and ammonium nitrate plants

3. **Ammonia, urea, phosphoric acid and complex fertilizer plants**
4. **Ammonia, urea, ammonium sulphate and sulphuric acid plants**
5. **Ammonia, urea, phosphoric acid, nitric acid, sulphuric acid and complex fertilizer plants.**

A wide variety of liquid-pollutant streams are discharged from different process units in a fertilizer industry. Ammonia and ure-containing liquid effluents come from ammonia and urea plants whereas, ammonium phosphate and ammonia-bearing wastes come from ammonium phosphate manufacturing units. Acidic wastes are from spillages in sulphuric acid plants as well as eluants formed during regeneration of cation exchant resin units. Eluants from anion exchange resins is alkaline in nature. There are a number of operations where the, scrubbing of gases is carried out by solutions containing sodium hydroxide, arsenic compounds, amines, potassium carbonate or even pure water, for a variety of purifications, recovery of materials, or pollution control itself. Water scrubbing after the gasification of naptha or other hydrocarbons (partial oxidation) gives carbon slurry. Cooling, tower blow-downs give chromium compounds which are very toxic in nature. A typical analysis of the combined liquid waste from a fertilizer complex giving out an average liquid effluent flow of 20,000 tonnes/day shows that, on an average, total nitrogen in the effluent is between 400 and 1000 mg/l, urea nitrogen between 400 and 800 mg/l, phosphates between 70 and 500 mg/l, fluoride (F) between 10 and 20 mg/l and arsenic between 1 and 2.5 mg/l, the chief gaseous pollutants are sulphur dioxide, NO_x, flu rine and particulate matter of different sizes and natures. The acceptable value of nitrogen in ammonia or urea form in water bodies is 1.2 mg/l, of fluorides (as F) 2 mg/l and of Arsenic (as As) 1 mg/l, individually or along with other toxic metal compounds.

Table 7.1 gives the nature, sources, concentration (average), effects and tolerance of limits pollutants from fertilizer industries.

TABLE 7.1

Nature, Sources, Concentrations, Adverse Effects and Tolerance Limits of Effluents from Fertilizer Industries

Pollutant	*Average concentration in the combined waste (mg/l)*	*Source*	*Pollution effects*	*Tolerance limits*
1	2	3	4	5
Ammonia	732 (as N)	Ammonia and urea plants	Toxicity to fish and aquatic life fish kills eutrophication.	1.2 mg/1 as N in receiving waters used for fish culture.
Urea	614 (as N)	Urea plant	Toxicit of ammonia after hydrolysis entrophication.	See limits for ammonia above.

(*Contd.*)

1	2	3	4	5
Arsenic	1.6 (as As)	Gas purification plant in ammonia and urea manufacture.	Causes black-foot disease accumulative poison, affects plants and crops.	1.0 mg/1 in industrial effluents individually or along with other metals
Oil	N.A.	Gas compressor houses in ammonia and urea manufacture.	Suppresses dissolution of oxygen in water, floats on water and forms ugly and iridescent slicks.	10 mg/1
Phosphate	77 (as PO_4)	Phosphoric acid and complex fertilizer units.	Along with ammonia causes algal blooms which increase cost of water treatment.	
Fluoride	15 (as F)	Scrubber effluents in phosphoric acid and super phosphate manufacture.	Causes dental and skeletal fluorosis, affects hatching of fish eggs.	2 mg/1 as F

Ammonia Plant Effluents

The main constituents from an ammonia plant are carbon particles, oil droplets, hydrogen sulphide and nitrogen-wash tail gases.

Removal of Carbon

Several fertilizer units use partial oxidation of naptha, fuel oil and other hydrocarbon feedstock for, the production of synthesis gas. During the cracking process, apart from gaseous products such of CO_2, CO and hydrogen, soot particles are also formed. Before purification and separation these gases, the soot particles have to be removed. This can be done by scrubbing the gases with water in venturi scrubbers. The resulting carbon, slurry is normally drained, causing serious problems. Choking of pipelines is a common occurrence. The fine carbon particles in water do not settle during primary treatment and can reduce bacterial activity in the secondary treatment. When let into river or sea water, they adversely affect aquatic life by blocking sunlight and oxygen supply. The very fine size of the carbon particles presents a number of problems in their separation. In one method the carbon slurry is mixed with mineral oil when the carbon particles go to the oil phase which is separated by gravity. Pelletization of the carbon and distillation of the oil helps recover carbon and oil. The carbon obtained by this method always contains a small quantity of mineral oil and can be used as fuel. Recently, vaccum filtration of the carbon slurry has resulted in almost complete recovery of carbon. The filtrate obtained contains less than 100 mg/l of carbon particles and can be reused for scrubbing gases. The carbon is obtained in the form of a wet cake which can be dried and used in applications such as paints, printing inks and rubber industries. The slurry, after scrubbing of gases from the gasification unit, is pumped into a rotary vacuum filter where the carbon cake is separated from the filtrate. The latter goes into the receiver provided with the necessary gravity leg and pump. The initial concentration of the carbon slurry is 1-2 per cent, and of the carbon cake about 20 per cent by volume.

Oil Removal

Another liquid stream coming from a typical ammonia plant contains oil from the compressor bay. An oil separator followed by adsorption by coke helps reduce the oil content to values lower than 10 mg/l.

Hydrogen Sulphide Removal

The feedstock for the production of synthesis gas consists of naptha, fuel oil and coke. As almost all feedstocks contain some quantity of sulphur, synthesis gas contains H_2S which is recovered and generally burnt. The SO_2 obtained from the combustion of H_2S can be converted to sulphuric acid where sulphuric acid plants exist. Alternatively, H_2S can be converted into sulphur by the Claus process.

Nitrogen-Wash Tail Gas

Nitrogen, available as a by-product from oxygen plants, is used in the purification step to remove carbon monoxide from synthesis gas. The exist nitrogen wash gas contains approximately 45 per cent by volume of Co, 5 per cent hydrogen, 5 per cent methane and the rest (45 per cent) nitrogen. In ammonia synthesis, methane is the main hydrocarbon impurity formed by the reaction.

$$CO + 3H_2 = CH_4 + H_2O$$

The exit stream has a fairly high calorific value (15-1800 kcal/kg) but the common practice, even today, is to flare this gas. The gas can be used (where nitric acid plants exist) for catalytic reduction of NO_x in the exit gases. This is being done in selected plants in India. Alternatively, the spent tail gas can be used as a low-heating-value gas for generation of low-pressure steam, or for heating purposes.

Ammonium Sulphate Plant

In plants producing ammonium sulphate using natural gypsum, or phospho gypsum, for every tonne of ammonium sulphate made, chalk produced as a byproduct is 0.75 tonne. For an average daily production of 1000 tonnes of ammonium sulphate, chalk produced is of the order of 750 tonnes/day. It contains impurities like ammonium hydroxide, ammonium sulphate and unreacted gypsum. After removing undesirable impurities followed by proper treatment and conditioning, the chalk can be utilized in the initial stages itself so that the large amounts of by products of ammonium sulphate plants can be fruitfully utilized.

Phosphoric Acid Plant

In phosphoric aeid plants there are solid, liquid as well as gaseous effluents.

Solid Effluent

The important solid product of phosphoric acid plants is gypsum and its quantity varies from 4 to 6 tonnes, tonne of P_2O_5 produced as phosphoric acid. As much as 500 tonnes of gypsum is produced daily by a plant of capacity 100 tonnes/day. The gypsum produced by the wet process (dihydrate-based process) contains fluorine ranging from 1 to 1.5 per cent and P_2O_5 from 1 to 2 per cent. Their presence makes its use undesirable in cement manufacture. Gypsum contains 92-94 per cent $CaSO_4.2H_2O$. The method of phosphoric acid manufacture has now been changed to the hemi-hydrate-dihydrate or dihydrate-hemihydrate process, making it possible to get pure grades of gypsum; containing low quantities of P25 and fluorine. Their contents are reduced to about; 0.13 and 0.04 per cent respectively in the new process. When the demand for gypsum is low, it can be converted into slurry form with water and stored in ponds, the overflow of which is neutralized and precipitated using lime solution from a slaking unit. The chief use of gypsum has been as a retarding agent in the; finished cement product, but recently its use as a soil conditioner has been tried out on a large scale, especially for alkaline soils, the small impurity of P_2O_5 also helps during soil; treatment, the phosphogypsum can also be used for the manufacture of sulphuric acid based on the Mueller-Kuehne process.

Liquid Effluent

The liquid effluent contains acidic and fluoride-bearing chemicals which can be neutralized and precipitated using lime solution from a slaking unit, the typical capacity of the slaker is 25 tonnes of lime/day resulting in a 10 per cent concentration of $Ca(OH)_2$ in the final solution. The lime; solution can be circulated to different acid plants and used in neutralization pits provided for the purpose.

Gaseous Effluent

The chief gaseous pollutant from phosphoric acid plants is fluorine. It has harmful effects on the health of living beings as it leads to bone decay and smarting of skin, the following are the stream concentrations allowed in existing and new plants.

Existing plants	60 mg/m3 at NTP
New plants coming up	25 mg/m3 at NTP
Single super phosphate plants :	
Existing plants	0.5 kg/tonne of product
New plants coming}p	0.1 kg/tonne of product
Triple super phosphate plants :	
Existing plants	0.3 kg/tonne of product
New plants coming up	0.075 kg/tonne of product

Earlier, tall chimeys were used to disperse the pollutants but they are not of much use these days due to the proximity of residential areas. The effluent gases containing fluorine can be scrubbed with silicic acid in high pressure venturi scrubber followed by a cyclone; separator; and mist collector. The recovered hydrofluoro-silicic acid is a starting material for a manufacture of cryolite and other fluorine-bearing compounds, the recovery of fluorine from gases from phosphoric acid plants is an important example; of pollution abatement combined with recovery and production; of important chemicals. Cryolite, used in the manufacture of aluminium, is still being imported in large quantities but can be manufactured from fluorine-bearing compounds wasted in enormous quantities even today. Emission of fluorine in gases can be brought down to 25 mg/m3 or approximately 0.2 kg fluorne/tonne of P_2O_5 as phosphoric acid. Two units in India are already recovering fluorine-based chemicals from gases.

The pollution control measures adopted in fertilize units at present are very minimal or incomplete. They include lime treatment of fluorides and phosphates, incorporation of DCDA technology for sulphuric acid plants and removal of soot particles during gasification of naptha or other hydro-carbons. More attention should be paid to the control of acid mist, NO_x fluorine and ammonia in gases and of urea/ammonia, chromium, arsenic, hydrocyanic acid, lime sludge, carbon/ash slurries and mineral wastes in liquid effluents. Control measures for some of them are at the development and implementation stage but are far from adequate. Although many fertilizer units have been in operation for a number of years, effluent treatment has not received the attention it deserves. For the treatment of fluorides and phosphates in liquids, quite a few fertilizer units have incorporated lime-treatment facility, quite efficiently yielding an effluent with 5 mg/l phosphates and 10 mg/l fluorides. The cost factor, both capital and operational, as also the disposal of sludges, have always presented problems. The newer fertilizer units are incorporating equipment such as API separators, ammonia strippers, urea hydrolysers, cyanide reactors, nitrification tanks and final clarifiers. Primary and secondary treatment is being provided for the treatment of industrial effluents mixed with sewage waste. In existing plants, the strategy adopted includes re-cycling and re-use of effluents wherever possible segregation of contaminated effluents and their treatment whenever possible and mixing the general mass of effluents thereafter so that the pollutants are diluted before disposal. Lime sludge and ash slurry are sent to burrow pits. Ammoniacal effluent is reduced to 500 ppm, from an original concentration of 2.5 per cent, in a steam stripper and stored in pits and aerated lagoons with huge capacities. The effluent from these pits can be taken to storm-water sumps and diluted with other contaminated effluents to bring down the concentration of ammonia. As dilution of the liquid waste is no solution to the problem of pollution it will be desirable to think of systems which help recover urea/ammonia from the nitrogenous waste in some form. Similar exercises should be carried out for arsenic, chromium and other gaseous pollutants.

Pollution Control in Petroleum Refineries and Petrochemical Units

Pollution due to petroleum refineries encompasses three stages, *viz.* crude-oil production, transportation and refining operations. During production, brines containing oil are discharged into seas or rivers. Oil spills during sea, rail and road transport lead to the pollution of seas, rail yards and road stations. The presence of tar balls on sandy beaches is the result of pollution of the seas, where large quantities of crude oil or its products are thrown during transportation of their ever-increasing shipments to different parts of the world. However, refinery operations where crude oil is actually processed are responsible for maximum water pollution. Here the crude oil is subjected to distillation, cracking and hydrocracking operations to give with naptha, kerosene, diesel oil, gasoline, fuel oil, lubricants and LPG, along with asphalt, coke, greases and waxes. A pipe-still sends the gaseous products to fractionating columns where light products like kerosene, light fuel oil and gasoline are recovered after processing. All streams of the atmospheric distillation unit are suitably processed to yield products in the desired proportions. Very large quantities of water are used in refinery operations. As much as 30-80 kg of water/l of crude oil processed is necessary for operations such as cooling where enormous amounts of heat of reactions and operations are removed. Of the total amount of water consumed, cooling operations account for more than 80 per cent. This leads to thermal pollution of water bodies, leading to increased biological activity combined with decreased solubility of oxygen in water, thereby adversely affecting the aquatic life. Table 7.2 gives the processes taking place in refining operations and the associated water pollution.

Oil-bearing liquid waste forms a thin oil layer over the surface of the receiving water bodies, cutting down sunlight and oxygen necessary for flora and fauna. It is also a nuisance aesthetically and impedes the photosynthesis of plants. It leads

TABLE 7.2

Refinery Processes and Liquid Waste

Refinery Process	*Liquid Waste*
Distillation	Effluent with ammonia, hydrogen, sulphide and phenols.
Desalting	Sodium chloride, phenols, sulphide or free hydrogen sulphide
Vacuum distillation	Phenol and oil
Naphtha hydrotreating	Sour condensate
Catalytic hydrotreating	Hydrogen sulphide, oil or phenols
Catalytic or thermal cracking	Phenols, hydrogen sulphide, ammonia, cyanides
Hydrocracking	Ammonia
Catalytic alkylation	Alkalies from washings and acids from drains.
Solvent processes	Solvents like phenol, sulphides, copper acetate, etc.
Treating processes	Organic and inorganic pollutants (sulphides, phenols) and emulsified oil.

to the fouling of beaches, sea resorts and boats affects adversely the performance of water-treatment plants and is a fire hazard. Phenol is not readily biodegradable, imparts bad odour and taste to water and is toxic to living beings. Hydrogen sulphide and other sulphur compounds in water emit obnoxious smells and can impart a black colour to receiving waters or corrode the process equipment including the conveying lines and treatment plant.

Characteristics of Liquid Effluent

The average characteristics of liquid effluent from a refinery are different from those from a petrochemical unit, they are summarized in the following sections.

Refineries

The average liquid-effluent characteristics of refinery units vary from refinery to refinery and process to process. An average Indian refinery has the following untreated waste characteristics :

Free oil	=2000-3000 mg/l
H_2S and sulphides	=10-220 mg/l
Phenols	=12-30 mg/l
Suspended solids	=200-400 mg/l
5 day BOD at 20 C	=100-300 mg/l
Alkalinity	=10=250 mg/l

Analysis of the raw waste from refineries in USA indicates that phenols are present to the extent of 50 mg/l, ammonia nitrogen 300 mg/l, dissolved solids 20(0 mg/l, sulphides 60 mg/l, free oil 60 mg/l, pH 8·5, COD 800 mg/l and suspended solids 80 mg/l The quantity and quality of liquid wastes differ from refinery depending on the size, nature of crude, amount of water used and efficiency of processing equipment. In general, phenols and their compounds account for 0.03-0.96 kg/tonne of crude handled and the COD in the waste 0.4-0.9 kg/tonne of crude handled.

Petrochemical Industries

A petrochemical-manufacturing unit may be a fertilizer complex or an organic-chemical complex. The liquid gaseous effluents emitted by a fertilizer industrial complex have already been discussed. The treated effluent is expected to meet the following standards.

pH	5.5-9.0
Suspended solids not to exceed	100 mg/l
Dissolved oxygen not less than	5 mg/l
BOD (5 days, 20°C) not to exceed	100 mg/l
COD not to exceed	1 mg/l

Bioessay test for fish shall show a 90 per cent survival in 96 h

Petrochemical industries are so diverse that it is impossible to generalize waste characteristics and arrive at any given set of treatment methods.

Depending upon the product mix, the effluents from different plants would contain fairly high concentrations of organic and inorganic pollutants. Each plant effluent requires a special study of the quality and quantity of liquid waste involved to arrive at a suitable technology for its treatment.

The following six general techniques can be followed to minimize pollution by a petrochemical unit.

1. Physical, biological and chemical treatment of liquid waste before disposal
2. Treatment and recycling of process water to reduce fresh-water requirement as well as pollution
3. Incorporation of process changes to keep the pollutant concentration at a lower level
4. Recovery of useful chemicals

5. Segregation of toxic-liquid wastes and removal of toxic substances
6. Equalization and proportioning of liquid wastes.

Refinery Liquid Waste-Treatment Methods

Preventive and remedial measures should be incorporated for the treatment of refinery liquid effluent streams.

Inplant Measures

These include reduction of oil leakage by proper maintenance of pumps pipe fittings and joints. The formation of oil-water emulsion should be treated separately by air flotation with or without the use of coagulating agents. The use of alum or activated silica as a coagulant and/or coagulant aid can reduce the oil content to values lower than 10 mg/l. Air flotation can remove oil with an efficiency of 60-90 per cent and suspended solids with an efficiency of about 75 per cent. The corresponding reduction in COD is 60-80 per cent. The use of acids, coalescence-aiding membrane structures and carbon-aluminium granular beds for the removal of free oil has been reported. The separators recommended by the American Petroleum Institute (API) are finding wide use for the removal of free oil. The principle of a rectangular separator is based on gravity settling of oil droplets under reduced velocity of phases, The depth-to-width ratio of the separator is between 0.3 and 0.5. The depth varies from 0.9 to 2.4 m and the width from 1.8 to 6 m. The dimensions of the unit are based on the average properties of water and oil droplets of average size 0.15 mm. The recommended horizontal linear velocity is 13.5 m/min. As the flow of liquids (inflow as well as outflow) through the separator is continuous, the positions of outflow of water and of the interface have to be arrived at depending on the hydrostatics. The API has given suitable dimension for devices such as skim pipes, baffles, etc. The efficiency of gravity separators is limited to about 60 per cent and the effluent from them may still contain 30-40 mg/l free oil. A tilted-plate interceptor based on a Shell design containing plate-packs at an angle of 45 is shown in Fig. 1. The oil droplets impinge on the inclined surfaces, coalesce, and collect on top. The oil-water separator are useful both from the viewpoint of removing floating oils close to the source (in plant measures) as well as being part of the integrated

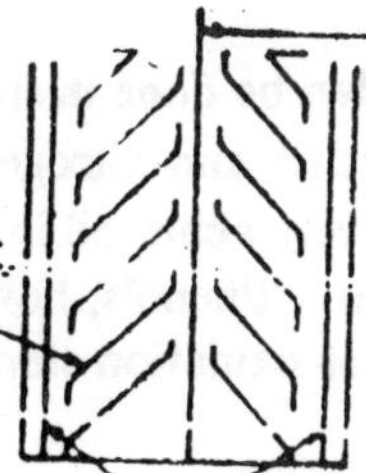

effluent-treatment system. Oil can usually be collected by the use of separators at each refining unit, at a number of conveniently-located sumps. The relatively-clear water is discharged into the main liquid-effluent stream while the trapped oil is sent for storage and use.

Good house-keeping practices such as eliminating leakages in pumps, discharge lines, as well as recycling of chemicals, process water and suitable disposal of solids go a long way towards reducing the pollution load and saving valuable chemical materials.

Steam Stripping

Sour-water condensates containing ammonia, hydrogen sulphide, mercaptans and phenols are produced during crude-oil distillation and catalytic and noncatalytic racking operations. The most widely-used method of treating liquid waste is by the use of sour-water strippers where hydrogen sulphide and ammonia are removed.

A proper mix of physico-chemical and biological treatments (for final polishing) would be desirable in this case. Air or steam stripping at higher pH reduces the ammonia content in water quite substantially.

Oxidation

Oxidation of sulphides can be done in spent caustic oxidizers. Oxidation of ammonia to water and nitrogen, and sulphides to thiosulphates, can be effectively achieved in columns where air is used as an oxidizing agent. Phenols, however, are left in the liquor during the stripping or oxidation steps.

Removal of Phenols

A multistage modified-activated-sludge process where phenol is reduced to 1 ppm has been reported. Adsorption of phenol on activated charcoal is particularly suited to refinery liquid wastes. It has been observed, recently that biological oxidation of phenols is possible provided specially cultured micro organisms and optimum conditions of temperature, dissolved oxygen, nutrients and pH are employed. For industrial practices, however, it is desirable that the chemical or physico-chemical processes discussed earlier are employed for the dephenolation of liquid wastes. Solvent extraction is ideally suited for recovering phenols from caustic cresylates and sour water from stripping towers. The cresylates recovered serve as cheap disinfectants.

Biological Treatment

Biological treatment can be used for treating/polishing the composite refinery liquid waste. For this use trikling filters, activated sludge, oxidation ponds, aerated lagoons and anaerobic digestation has been recommended. The general experience of dsigners has been that deep trickling beds are better suited for biological oxidation as compared to the activated sludge process for refinery wastes where traces of oil are present. Retention times have also been found to be higher if slight amount of phenols are present in the liquid waste. Oils, if present, also tend to increase the time of treatment. At present, the general trend is to treat only some selected dilute streams containing phenol and with little or no oil content. Ideally, however, one should treat all contaminated streams or the total waste containing all pollutants. The maximum efficiencies for biological treatment using oxidation ponds, trickling filters, aerated lagoons and activated-sludge process varies from 80 to 95 per cent respectively whereas the minimum efficiency for BOD removal varies from 4 to 70 per cent. Several modifications in the conventional activated-sludge-process have been reported for treating liquid wastes from petroleum-refining and some petrochemical-processing units. They include multistage activated sludge treatment, ferrobiological process (with ferric oxide flocs) and UNOX process where pure oxygen is employed instead of air. A number of treatment plants have been installed in oil refineries in eastern and western part of India for the treatment of integral effluents containing chiefly oils, sulphides, phenols, BOD and suspended solids. These treatment plants include API separators for the removal of oil, chemical treatment and clariflocculators for the remeval of emulsified oils, sulphides and suspended solids and biological treatment employing deep tickling filters/activated-sludge process for the removal of phenol and BOD. In some plants a coke-bed separator has been employed for the removal of fine carbon particles. The final oil content from the integrated treatment plant is as low as 5 mg/l. Other pollutants are also removed to the desired levels. It is claimed that phenol and BOD concentration can be reduced to less that 1 mg/l and 30 mg/l respectively as is usually required.

Disposal of Sludges

Sludges are formed during oil-water separations, precipitation operations and biological oxidation. They are essentially of three types-oily sludges chemical sludges and biological sludges. The only sludges can be nickened by gravity thickeners, centrifuges orfilteration. They can shen be dried and burnt or disposed off as land fills. The chemical sludges forme during chemical as well as oil processing can be used for the recovery of suitable materials or used as they are depending on their types. Calcium sulphate and calcium carbonate sludges can be processed for cement manufacture or used as land fills along with acidic sludges. Sulphuric acid alkylation sludge, calcium fluoride from hydrogen fluoride alkylation, acidic sludge from the treatment of lube oil stocks and spent aluminium chloride complexes are drained free of water and buried with chalk or other suitable alkaline solid wastes. Biological sludges obtained from biological oxidation should be digested anaerobically for the production of biogas, unfortunately, in India little or no effort is being made for suitable disposing of the sludges from treated liquid effluents. The accumulating heaps of sludges in and around the factory area and paucity space where sludges and treated liquids can be thrown without causing environmental problems bear testimony to the lack of desire and efforts to disinfect or otherwise render innocuous these sludges or treated liquids.

TREATMENT OF LIQUID EFFLUETS FROM PETROCHEMICAL INDUSTRIES

The treatment of liquid effluents from petrochemical industries can be done on the same lines as refinery wastes.

The concentration of oils, phenols, ammonia and compounds specific to the chemical plant should first be reduced or recovery by phsico-chemic l methods. A survey of techical literature published on different processes for treatment of liquid waste after recovery of various chemicals present indicates that the processes involved should be an optimum combination of the following :

1. Solvent extraction
2. Inertial separation
3. Ion exchange
4. Adsorption on activated carbon or polymeric resins

5. Recovery of useful chemical
6. Chlorination
7. Electrocoagulation
8. Electrodialysis
9. Ozonation
10. Biodegradation

There are a number of instances where useful compounds have been recovered from wastes from individual plants in a petrochemical complex or other chemical-process industries. Resorcinol has been recovered from a meta-aminophenol plant, di-nitrocresols from nitrotoluene effluents KCI from a pharmaceutical industry and high-grade $CaCO_3$ from $CaCl_2$ waste from a pickling unit, to name a few examples. Wherever porsible, solvent extraction should be used for the recovery of useful organic substances miscible with water. The floating oils and immiscible organics should be separated with an API separotor. Ion exchange and dialysis or other membrane techniques help to recover the materials and should be preferred for high-cost chemicals. Facilities for neutralization, clariflocculation and thickening also go a long way towards treasing liquid wastes after useful materials have been recovered. The recycling of water in the processes is a consideration which needs emphasis. Segregation and regrouping of different effluent streams, depending on the nature of the recovery or treatment envisaged, deserve attention.

Unfortunately, not much attention is paid to the treatment of liquid wastes from petrochemical industries as these are termed difficult wastes. However, scientific rational insight would reveal that the recovery/treatment methods are not greatly different from the actual processes taking place in a petrochemical unit. At best, they are a little more intensive but technically manageable. Studies on carbon adsorption, for example, indicate that a majority of indigeneously available carbons are poor in quality. Imported varieties are also specific to some of the pollutants such as phenols, but a majority of polar chemicals, such as nitroaromatics are not effectively adsorbed There is great scope for development in the field of active-carbon production for specific applications. Polymeric adsorbents have been found to be effective because of the low binding energies involved.

Besides, their regeneration characteristics are also much better. Polymeric adsorbents on a base such as palystyrene should also be made available for applications envisaged in the future. Chlorination and ozonation are known to be effective in the destruction of organics such as phenols but the chlorine-to-phenol ratios required are very high, ranging from 5 to 10. They can be employed for a final polishing of the wastes after the adsorption step. Biological oxidation follows other methods in an integrated treatment plant and works efficiently only in the removal of residual BOD after most toxic materials have been recovered or removed.

8

Prediction and Assessment of Impacts on the Water Environment

Introduction

One of the major impacts from many actions is evidenced by changes in water quality both in the vicinity and downstream from the project areas. Construction of reservoirs, power plants, industrial parks, and pipelines will cause short-term impacts on the water environment : and operation of these same facilities will result in longer-term impacts. General impacts on the water environment are related to hydraulic and hydrologic cycle changes as well as to the introduction of suspended and dissolved materials into receiving waters. This lesson addresses the data needs and associated technology for predicting and assessing impacts of proposed actions on water quality.

Basic Steps for Prediction and Assessment

Basic steps associated with prediction of changes in the water environment and assessment of the impact of these changes are as follows :

1. Determine types and quantities of water pollutants emitted from all alternatives for meeting a given need during both construction and operational phases.

2. Determine the existing water quantity and quatily levels for the surface watercourses in the area. Examine the frequency distributions and the median and mean data for both water quantity and quality. If possible, consider historical trends of water quality. Note particularly the low flow ultilized by the local regulatory agency for maintenance of water quality standards.

3. Document unique pollution problems that have occurred or are existing in local surface watercourses.

4. If relevant for the project alternatives, describe groundwater quantity and quality in the area, noting the depth of the groundwater table and direction of groundwater flow. Identify major local uses of groundwater, and delineate historical trends for groundwater depletion and pollution.

5. Assemble summary of key meteorological parameters for the area, noting particularly the monthly averages of precipitation, evaporation and temperature.

6. Procure the applicable water quality standards for local surface watercourses and groundwater supplies if relevant. Specify applicability of effluent standards and required treatment technology and state whether the receiving stream is water-quality limited or effluent limited. Consider time schedules required for attaining applicable water quality standards.

7. Summarize the organic waste load allocation study for area. Also procure extant information on inorganic, thermal, sediment, and bacterial waste loads. Identify known point sources of pollution, focusing specifically on unique discharges or wastewater constituents. Also enumerate the types of water uses in the area and summarize the quantities involved.

8. Determine the mesoscale impacts by calculating estimated daily quantities of water pollutants from the alternatives during both construction and operational phases and comparing these to existing waste loads in

the drain area. Determine the percentage increase in these waste loads. Note existing water quality parameters that are good or poor relative to current or potential standards.

9. Consider construction phase impacts in terms of the following factors :

   (a) Time period of construction and the resultant time period of decreased water quality. Specify stream discharges and quality variations that would be anticipated during the construction phase.

   (b) Anticipated distance downstream of decreased water quality.

   (c) Implications of decreased water quality relative to downstream water users. If there are uses that require certain water qualities, identify the required raw water quality characteristics and discuss the effects of decreased quality during the construction phase.

   (d) Specific construction specifications directed toward pollutant minimization.

10. Determine the microscale impacts by calculating specific downstream concentrations resulting from conservative pollutants, dissolved oxygen concentrations resulting from nonconservative (organic) pollutants, and temperatures resulting from thermal discharges consider these microscale impacts for both construction and operational phases. Compare calculated downstream concentrations with applicable water quality standards. Check if applicable effluent standards are met for existing facilities, or consider how they will be brought into compliance. In the case of new sources. Identify necessary technology for compliance with new source, performance standards.

11. If water quality or effluent standards are exceeded, consider mitigation or control measures.

12. Consider operational impacts of alternatives in terms of the following factors :
    (a) Frequency distribution of decreased quality and quantity.
    (b) Effects of sedimentation on the stream bottom ecosystem.
    (c) Fate of nutrients by incorporation into biomass.
    (d) Reconcentration of metals, pesticides, or radionuclides into the food web.
    (e) Chemical precipitation or oxidation/reduction of inorganic chemicals.
    (f) Anticipated distance downstream of decreased water quality and the implications for water users and related raw water quality requirements.
    (g) General effects of any water quality changes on the stream ecosystem.
    (h) Unique water quality changes that occur as a result of water impoundment and thermal stratification.

These twelve steps are directed toward determining the water impacts of alternatives on the mesoscale levels. Mesoscale assessment considers the contribution of alternatives to area water pollutant sources, both point and non-point. Microscale assessment involves comparisons of calculated concentrations of water pollutants to applicable water quality standards. Both levels of impact assessment are necessary in order to adequately address water quality impacts associated with proposed actions.

Steps 9 and 12 are directed toward a general discussion of the features and ramifications of construction phase and operational phase impacts. Step 12 is suggestive of the types of factors that should be considered under operational phase impacts, but this list is not intended to be all inclusive of items that should be addressed. Detailed coverage responsive to step

12 should be related to the type of project and the environmental setting. Steps 2-7 should be summarized in the environmental setting section of the impact statement, with the remaining steps included in the environmental impact section.

The organization of this lesson is primarily based on the twelve steps identified above. However, prior to a step by-step discussion and analysis, brief information on basic water pollution considerations is presented.

Basic Information on Water Pollution

Water of high quality is essential to human life, and water of acceptable quality is essential for agricultural, industrial, domestic and commercial uses, in addition, most recreation is waterbased; therefore, major activities having potential effects on surface water are certain to be of appreciable concern to the consumers and taxpayers who support these activities. Additionally, developments of recent years suggest that now people are far more concerned about water quality than in previous years.

Water pollution can be defined in a number of ways; however, the basic elements of most definitions are the concentrations of particular pollutants in water for sufficient periods of time to cause certain effects. If the effects are health related, such as those caused by pathogenic bacterial intrusion, the term "contamination" is appropriate. Effects that have to do with limitations on water availability due to certain water quality requirements related to usage can serve as a basis for defining a condition of water pollution. "Nuisance" refers to aesthetically displeasing effects created by oils, grease, or other floating materials.

Potential impacts on surface water quality and quantity are certain to be of concern in assessment of the effects of many development programs. Almost any activity of man offers the potential for impact on surface water through generation of waterborne wastes, alteration of the quantity and/or quality of surface runoff, direct alteration of the water body, modification of the exchanges between surface and groundwaters through direct or indirect consumption of surface water or other causes.

The hydrologic environment is composed of two interrelated phases : groundwater or surface water. Impacts initiated in one phase eventually affect the other. For example, a groundwater system may charge one surface system and later be recharged by another surface water system. The complete assessment of an impact dictates consideration of both groundwater and surface water. Thus, pollution at one point in the system can be passed throughout, and consideration of only one phase does not characterize the entire problem.

Due to the close interrelationship between surface and groundwaters, most environmental attributes inevitably interface. Hence, aside from those aspects dealing specifically with surface or subterrain features, the attributes may be considered as applicable to both.

Potential water quality impacts must be considered based on a clear delineation of various water quality characteristics. It is necessary to utilize a manifold evaluation of water quality characteristics in order to develop a total evaluation of existing water quality as well as microscale changes that might result from project alternatives. Water quality can be described in terms of physical, chemical, and bacteriological attributes.

Physical attributes of surface water could be categorized as relating to either the physical nature of the water body or to physical properties of the water contained therein.

Physical parameters include color, odor, temperature, solids (residues), oils, and grease. Color can be defined relative to type and density, the type being related to whether it is true color (dissolved) or apparent color (filterable). Odor is described by type and threshold odor number, which is related to the odor-free water required for diluting an odorous water sample to a nonodorous level. Figure 1 shows the relationships between various solids parameters used in water quality characterization. Total solids are comprised of suspended and dissolved solids, and each of these fractions can be further divided into organic (volatile) and inorganic (fixed) components. Turbidity is another measure of the solids content, and it is related to light transmittance through water. Settleable solids describe the materials present in solution that will settle by gravity in a 1-hr period. Specific conductance (conductivity) is a measure of the inorganic dissolved solids present in ionic form. In surface watercourses oil and grease is of interest relative to nuisance considerations.

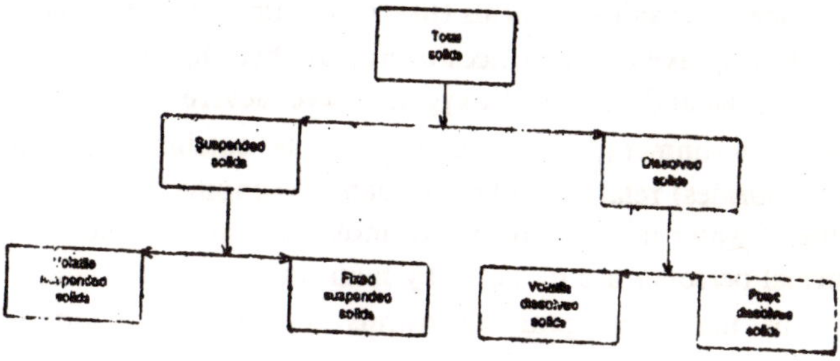

Fig. 1. Relationships of various solids tests used or water quality characterization.

Many types of activities could influence the physical properties of water. A few examples are clearing of land and construction of hardstands, roads and rooftops (which might accelerate erosion, flooding, and sedimentation), discharge of of scale-laden boiler waters, and discharge of cooling waters. Some other quality aspects which could be included in this category are dissolved gases and tastes and odors, which are actually manifestations of chemical properties of water. This serves to illustrate the occasional difficulty in categorizing attributes of the water environment. Some other parameters which may be included in this category are the depth, velocity, and rate of discharge of a stream. Features of this type might be influenced by major activities, such as withdrawal of water, dredging, and clearing of vegetation.

Chemical attributes could be categorized conveniently as organic or inorganic chemicals. Inorganic parameters of potential interest in water quality characterization include salinity, hardness, pH, acidity, alkalinity, and content of iron, manganese, chlorides, sulfates, sulfides, heavy metals (Hg, Pb, Cr, Cu, Zn), nitrogen (organic, ammonia, nitrite, nitrate), and phosphorous. Salinity and chloride content are a measure of the salt in water. Hardness is caused primarily by divalent metallic cations that have soap-consuming potential, the major ones being calcium and magnesium. Nitrogen and phosphorus contents are of interest due to their nutrient characteristics.

Some inorganic chemicals (like cadmium, lead, and mercury) may have grave consequences to human health; some (notably phosphorus and dissolved oxygen, have severe effects on the water environment, while others (such as calcium, manganese, and chlorides) relate mainly to man's economic and aesthetic value of water in his commercial, industrial, and domestic uses. Normal personal use of water by man increases the concentration of many inorganic chemicals in water. Additionally, almost any type of industrial activity and land drainage is a source. Because of the hundreds of thousands of organic (corbon-based) chemical category are "lumped-parameters". Examples include BOD (biochemical oxygen demand), oil, and toxic compounds. Some organic compounds are natural constituents of surface drainage and human and animal wastes, while others are unique to industrial activities and industrial products.

Several tests can be employed to describe the organic characteristics of water. The most-used test is the biochemical oxygen demand (BOD), which is defined as the amount of oxygen required by bacteria in decomposing organic material in a sample under aerobic conditions at 20°C over a 5-day incubation period. Other tests that describe the organic content of water include the chemical oxygen demand, total organic carbon, and total oxygen demand.

presence of pathogens. Specific pathogens such as salmonella organisms may be relevant for certain environmental impact studies. Minimal information is available on virus types and concentrations in surface waters and pollutant discharges therein. Pathogenic (disease-causing) agents include certain virus, bacteria, protozoa, and other organisms, and they originate almost exclusively from human wastes. Aquatic life refers to the microorganisms and microscopic plants and animals, including fish, which inhibit water bodies. They are affected directly or indirectly by almost any natural or man made change in a water body.

There are two main sources of water pollutants in surface watercourses, namely, point sources and non-point sources. The total waste load in a stream is represented by the sum of all point and non-point pollutant sources.

Individual sources of water pollutants in aqueous systems are manifold, and, resultant water quality probably will not be attributable to single types of pollutants or single sources. Some of the major sources of water pollution are municipal wastes, agricultural wastes, solid erosion, accidental spillage of oil and other hazardous substances, acid mine drainage, mine sediments and watercraft waste.

Waste waters from municipalities are increasing due to expanding populations and greater per capita water usage. Waste loads from municipalities are expected to quadruple in the next 51 years, and over 1,000 communities outgrow their treatment facilities each year. It is estimated 50 years, and over 1,000 communities outgrow their treatment facilities each year. It is estimated that only one-third of the populations is served with adequate sewers and waste-water treatment facilities. In addition, storm and combined sewers contain a collage of pollutant materials washed from streets and urban areas.

The growth rate of industrial waste-water discharges is greater than that for municipal discharges. Quality characteristics of industrial wastes vary considerably depending upon the type of industry. A useful parameter in describing industrial waste is population equivalent.

Agricultural wastes include irrigation return flows as well as runoff from feedlots. These wastes exhibit salinities that are severalfold greater than unused irrigation water ; also, hardness, total dissolved solids, and turbidity are at increased levels, Irrigation return flows may also exhibit increases in nitrogen, phosphorus, and pestieide contents. Runoff waters from feedlots have high organic, nutrient, and solids concentrations and contain nicroorganisms that are potentially pathogenic to animals and humans. As an illustration of the organic strength, a feedlot with 10,000 cattle produces runoff with a BOD content equivalent to the untreated waste water from a city of 45,000 persons.

Soil erosion is one of the major water pollutants in terms of quantity. The total quantity of solids faom soil erosion is approximately 700 times greater than the totai from municipal waste-water discharges.

Accidental spillage of oil and other hazardous substances into watercourses is of fairly recent interest. Events such as the grounding of the Torrey Conyon in 1967 and the Santa Barbara offshore oil leak in 1969 have dramatized that these occurrences can cause devastating and extensive damage to the aqueous environment.

Acid mine drainage is of particular importance. Ptoblems are genereted when water and air react with sulfur-bearing minerals in mines or refuse piles to form sulfuric acid. Acid mine drainage also contains iron, copper, lead, zinc, and other metals that may be toxic to aquatic life. Sediment yields from strip-mining operations are also important *e g*. 230 mi^2/yr are disturbed by surface mining operations in the total United States, and sediment yields may range up to 30,000 tons/mi^2-yr from strip-mine areas. Watercrafts also discharge sanitary wastes, oils, ballast, nnd bilge waters. Although the total quantities of waste discharged from watercraft are small relative to other pollutant sources, they are important since most of the discharges are in high-use shoreline and harbor areas.

The effects of water pollutants on receiving water quality are manifold and dependent upon the type and concentrations of pollutants. Soluble organics, as represented by high BOD wastes, cause depletion of oxygen. Trace quantities of certain organics cause undesirable tastes and odrs, and some may be biomagnified in the food web. Suspended solids decrease water clarity and hinder photosynthetic processes; if solids settle and form sludge deposits, changes in benthic ecosystems result. Color, turbidity, oils and floating materials are of concern due to their aesthetic undesirability and possible influence on water clarity and photosynthetic processes. Excessive nitrogen and phosphorus can lead to algal overgrowth with concomitant water treatment processes. Chlorides cause a salty taste to be imparted to water, and in sufficient concentration limitations on water usage can occur. Acids, alkalis, and toxic substances have the potential for causing fish kills and creating other imbalances in stream ecosystems. Thermal discharges can also cause imbalances as well as reductions in stream waste assimilative capacity. Stratified flows from thermal discharges minimize normal patterns in receiving streams and reservoirs.

STEP 1 : Identification of Water Pollutants

The first step in prediction and assessment of water quality impacts involves identification of types and quantities of water pollutants emitted from construction and operation of each alternative under consideration. One approach for identifying water pollutants is to review EISs prepared for similar projects. Another approach is to utilize the unit waste generation factor, which is defined as the rate at which a pollutant is released to a drainage area or watercourse as a result of some activity, such as land clearing or production by industry, divided by that activity.

It is difficult to conceive of an alteration of surface water quantity or quality which is not accompanied by secondary effects. The physical, biological, and chemical factors influencing water quality are so interrelated that a change in any water quality parameter triggers other changes in a complex network of interrelated variables. Thus, while individual water quality and quality parameters may seem far more amenable to quantitative expression than parameters describing the quality of other sectors of the environment, the total effect of a particular impact on surface water may be as intangible as those on any sector of the total environment, because of the complex secondary, tertiary, and higher order effects.

STEP 2 : Description of Existing Water Quantity and Quality Levels

The second step involves assembling information on existing water quantity and quality levels in the area of the project, particularly focusing on quality parameters related to anticipated water pollutants to be emitted from construction and operational phases of the project. Sources of water quality informa-include relevant city, state water resources agencies and private industries that have monitoring programs. Another source is the STORET system of the EPA, with STORET meaning storage and retrieval of water quality data.

Since water quality standards vary with the beneficial uses assigned for particular streams or stream segments, it may be necessary to evaluate existing water quality relative to various standards. This step is important for projects that may have an impact over large distances in a single stream and for other projects, such as pipelines, that may cross numerous streams and several states. Interpretation of water quality data involves comparison with water quality standards. A rating system that has been used is based on assignment of zero values for parameters in compliance with applicable standards. Table 8.1 includes a list of numerical assignments used for this stream quality rating system.

TABLE 8.1

Numerical Assignments for Stream Quality Rating System

No. of Samples	Value	Physical	Chemical	Bacteriological
		Numerical assignments for parameter classification		
<10	Max.	−1	−2	−3
	Min.	−1	−2	−3
	Mean	−3	−6	−9
>10	Max.	−2	−4	−6
	Min.	−2	−4	−6
	Mean	−6	−12	−18

A meets all existing standards; excellent water quality (score=10)

B-meets most existing standards; good water quality (score=−1 to −10)

C-Does not meet few existing standards; fair quality (score=−11 to −30)

D-Does not most several existing standards; poor water quality (score= - 31 and greater)

The rationale for Table 8.1 was that not meeting a bacteriological standard was three times as important as not meeting a physical standard, and not meeting a chemical standard was twice as important as not meeting, a physical standard therefore, the negative numerical assignments for the bacteriological classification were three times greater than those for the physical classification, and the chemical classification assignments were two times greater than those for the physical classifications. In addition, when the mean value of a parameter did not meet the applicable standard, it was considered three times as important as when the maximum or minimum did not meet the standard; therefore, the negative numerical assignments for the mean valves were three times greater than those for the maximum or minimum values. Finally, stations from which 10 or more samples have been analyzed are twice as important as those stations from which fewer than 10 samples have been analyzed; therefore, the negative numerical assignments for stations with 10 or more samples were twice as great as the comparable assignments for stations with less than 10 samples.

Determination of a numerical score for a station in the study was based on a composite summation of negative numerical assignments applied when standards were not met. The zero or negative scores for each station were assigned a rating on the following basis :

Stream flow information can be obtained from the Indian Geological Survey as well as from local and state agencies dealing with water resources.

One of the key concerns with regard to stream flow is the flow frequency, which is utilized for calculations of compliance with water quality standards. In some instances the 7 day, 2-year low flow is utilized, in other cases 7-day, 10 year low flow is required. The phrase 7 day, 2-year low flow indicates that this is the minimum flow that occurs over a 7-day period on a frequeney of once every 2-years. Flow frequency information is available from local and state water resources agencies.

STEP 3 : Unique Pollution Problems

The primary purpose of this step is to identify any unique pollution problems that have occurred in the project area. This is necessary in order to adequately describe the environmental setting, to indicate a familiarity with the area, and to focus on environmentally sensitive parameters. Examples of unique pollution problems that should be identified included fish kills, excessive algal growth, and thermal discharges causing stratified flow. Many sources can be used to obtain information on unique pollution problems. Local and state water resource agencies constitute one source, and conservation groups, another. Local newspapers provide historical documentation of pollution concerns.

STEP 4 : Description of Groundwater Quantity and Quality

This particular step may not be required for all project types, but for alternatives that have potential for groundwater impact, this is a necessary step. This basic purpose is to determine the depth to the groundwater table in the area and to identify the direction of groundwater from the area should be enumerated, as well as historical trends in groundwater depletion for quality deterioration. Sources of information include local and state agencies dealing with water resources and the Indian Geological Survey, Central Soil and Water Research Institute etc.

STEP 5 : Summary of Meterological Information

Meterological data are required in order to predict and assess air quality impacts associated with proposed actions. In addition, certain climatological factors such as precipitation, evaporation, and air temperature are important in terms of predicting and assessingh water quality impact. In addition, precipitation and temperature information may need to be considered for proper construction scheduling. The primary sources of information include local and state water resources agencies as well as Oceanographic Institute of Indian and Indian Meterological departments.

STEP 6 : Water Quality Standards

The Water (Prevention and Control of Pollution) Act deals with only one aspect of the environment, *viz.*, water, as we do not have one comprehensive act for control of all types of environmental pollution. The object of the law is to control pollution of water and to maintain and restore, whenever necessary, the wholesomeness of water. The purpose of the law is not to restore water to its prestine quality but to make the best use of water for development activities and still preserve its wholesomeness for other uses. Programmes have, therefore, to be evolved for maintaining and restoring, the quality of water for its best use, as well as to determine the degree of pollution control necesssary before waste water is let out The central and state water pollution-control boards adopted the following use-based classification of waters :

1. Fresh Waters

Classification	*Best use to which it can be put*
A	Drinking Water source without treatment but after disinfection
B	Outdoor bathing
C	Drinking Water source with conventional treatment followed by disinfection
D	Propagation of wild life fisheries
E	Irrigation, industrial cooling and controlled waste disposal

2. Sea Waters (Including Estuaries and Tidal Waters)

Classification	*Best use to which it can be put*
A	Water sport (contact), shell fishing, salt pans
B	Commercial fishing, non-contact recreation
C	Industrial-cooling
D	Harbour
E	Navigation, controlled waste disposal

No separate classification exists for water used in industrial processes as its quality requirement varies from industry to industry. Of all the uses of water the one for domestic consumption is of prime importance. Water-borne diseases such as infections hepatitis : dysentry, cholera and typhold are caused by the contamination of water by domestic-waste discharges. The prevention of pollution of drinking-water sources in India, by the enforcement of suitable preventive measures, should have the highest priority. This calls for prohibition of discharges, both treated and untreated, into water bodies used for drinking and bathing purposes. The standards for drinking water are listed in Table 8.2. The MPN index of coliform bacteria should be less than one and none of the samples should have an index exceeding 10/100 ml. The major sources of pollution of fresh and coastal waters are the community wastes generated by cities and towns where the population is more concentrated. In the Class I cities of India where 10 per cent of the century lives, about two-thirds of tke country's pollution, responsible for the deterioration of the water quality of rivers and streams, is generated. This pyllution renders the water unfit for drinking purposes, especially in villages where the population is exposed to water-borne diseases. Only about one third of the Class I towns have been provided with sewage treatment facilities. Other towns in India where the population is relatively smaller, do not, in many cases, have sewage collection and/or treatment facilities. The main reason for this state of affairs is the paucity of funds with local bodies. It becomes necessary, therefore, for the state, central and other funding agencies to provide adequate financial help to local bodies so the latter can incorporate the necessary collection, treatment and disposal facilities. The provision of sewage treatment facilities in urban areas and towns has the following distinct advantages.

TABLE 8.2

Standards for Drinking Water

Pollutants	*Permissive*	*Excessive*
1	2	3
Physical		
Turbidity (units or silica scale)	8 units	25 units
Colour (units on platinum-cobalt scale)	5 units	25 units

(Contd.)

1	2	3
Taste and odour	Nothing	Disagreeable
Chemical		
pH	7.0.8.5	Less than 6.5 or greater than 9.2
Total solids	500 mg/l	1500 mg/l
Total hardness (as $CaCO_3$)	300 mg/l	600 mg/l
Calcium (as Ca)	75 mg/l	200 mg/l
Magnesium (as Mg)	50 mg/l	150 mg/l
Iron (as Fe)	0.3 mg/l	1.0 mg l
Manganese (as Mn)	0.1 mg/l	0.5 mg/l
Copper (as Cu)	1.0 mg/l	3.0 mg/l
Zinc (as Zn)	5.0 mg/l	15.0 mg/l
Chlorides (as Cl)	250 mg/l	1000 mg/l
Sulphates (as SO_1)	250 mg/l	400 mg/l
Phonolic substances (as Phenol)	0.0001 mg/l 1.0 mg/l	0.002 mg/l
Fluoride (as F)	1.0 mg /l	2.0 mg/l
Nitrates (as NO_2)	20 mg/l	50 mg/l
Taxis Substances**		
Arsonic (as As)	0.2 mg/l	0.2 mg/l
Chromium (as hexavalent)		0.05 mg/l
Cyanides (as CN)		0.01 mg/l
Lead (as Pb)		0.3 mg/l
Solenium (as Se)		0.05 mg/l
Radio Activity		
Alpha emitters (uc/ml)		10^9
Beta emitters (uc/ml)		10^8

*Figures in excess of the permissive, while not acceptable may still be tolerated in the absence of alternative and better sources but up to the limits destiguated above which the supply will not normally be acceptable.

**The values shown are maximum values which should not be exceeded.

1. Industrial liquid effluents can be treated at a small extra cost.
2. Downstream water quality will improve.
3. General sanitary conditions in the towns will improve.
4. Reuse of water can be planned in a big way, especially for agriculture.
5. Financial help from industries towards operating costs would be forthcoming.

The water pollution control act has provisions only for controlling pollution caused by industries or communities. It provides only for prohibitive and restrictive measures and does not empower the regulating agencies to instruct development authorities to reduce pollution. The regulating agencies are responsible only for qualitative aspects of pollution, whereas quantitative aspects are left out. This gives very restricted scope for the development of strategies and policies for achieving the objectives of the act. Therefore, only short-term measures for individual pollution problems can be dealt with.

The water quality-management objectives can be met by four general approaches :

1. Uniform effluent-standards approach.
2. Using assimilative capacity of receiving water approach.
3. Technology-based effluent standards.
4. Minimum national-standards approach.

Prescribing uniform standards is simple and straight forward. It is easy to estimate and enforce since the requirements are clearly defined. There are, however, some limitations which must be kept in mind. Uniform standards are not equitable as the volume of the waste is not considered. Quality standards alone are not sufficient for defining the requirements of effective pollution control.

Using receiving water quality as controlling factor is most appropriate as the maintenance of this is the objective of the legislation, unfortunately regulating bodies find this approach extremely difficult to enforce as they are required to regulate multiple discharges and apportion the responsibility. Technology-based effluent standard approach is in practice in USA and the first step in this is the adoption of what is called the 'Best Available Technology Economically Achievable' (BATEA) approach. This approach requires categorization of industries based on products manufactured and setting up of uniform national standards for each category after a detailed study of the possible modes of liquid treatment and the associated costs. This is a good approach and should be followed, especially for industrial-pollution control. There is a need for adopting similar standards for industrial waste water but it will be long before such an exercise is complete. The chief limitation of this approach is that the standards are more or less subjective and may have to be updated with continuous updating of technology.

It is therefore, felt to be logical to adopt an approach which requires a minimum treatment of effluents to maintain the stream standards for the designated use. The central board in India has evolved certain minimum industry-specific effluent standards at the national level which are designated the Minimum National Standards (MINAS). This would envisage certain minimum standards in the treatment of all wastes regardless of location. No relaxation will be permitted, but if the quality criteriou at a certain location so warrants, the state boards can after the standards. This approach is expected to be quite useful in halting the immediate threat of water and air polution as well as providing a measure of flexibility to the state Boards in the control of pollution. For MINAS, the water treatment envisaged should include the removal of pathogens (by effective disinfection), toxic materials, mineral oils, dissolved organic solids, suspended materials and a proper control of pH.

Summarizing, it may be noted that the existing water control act has provisions for taking only restrictive and prohibitive measures. There are no provisions for controlling the developmental activities which may effect water quality. Both central and state boards are unable to execute actual projects required for the prevention and control of pollution. Granting consent letters and launching legal action against the defaulters is hardly adequate. Lack of funds with local bodies, and proper incentives/deterrents for industries to control pollution is another limitations.

STEP : Waste Load Allocation Study

The purpose of this step is to summarize the waste load allocation study for the particular surface water-cources in the vicinity of proposed alternatives. Most states are developing or have recently completed studies of waste loads in surface waters. One result of this step is the identification of known point sources of pollution in the vicinity of the study area. Attention should also be directed toward unique discharges or effluent constituents in the area.

It is also important to identify numbers and types of water users in the area, particularly those downstream from the project site. Water quantity concerns are of major importance in water deficient areas. The type of water uses are important since quality requirements vary for different uses.

STEP 8 : Mesoscale Impact Calculation

The purpose of this step is to examine the impact of alternatives in terms of their relative contributions to existing waste loads in streams. The approach consists of multiplying unit waste generation factors by their appropriate production quantities and then comparing these calculated waste loads with existing waste loads in the study area. One means of assessing the impact is to calculate percentage changes in pullutant loads resulting from the alternatives relative to existing pollutant loads in the study area. Waste loads should be considered for organic (nonconservative), inorganic (conservative), solid, nutrient, bacterial and thermal pollutants.

STEP 9 : Construction Phase Impacts

The primary water quality impact during construction results from sediment that is eroded from the construction site, transported to local surface watercourses, and then dispersed or deposited. Many predictive methods have been developed, *e.g.* Lacey's Silt theory, Empirical formulas and some statistical formulas.

STEP 10 : Microscale Impact Calculation

The next step involves calculation of downstream concentrations of various water pollutants from the project area for each alternative during both the construction and operational phases. It is beyond the scope of this presentation to derive various mathematical formulas used for microscale impact prediction; however, this section will be oriented to presentation of several mathematical models that can be used in microscale impact calculations.

Conservative Pollutants

Conservative pollutants are not biologically degraded in a stream, nor will they be lost from the water phase due to precipitation, sedimentation, or volatilization. The basic approach for prediction of downstream concentrations of conservative pollutant is to consider the dilution capacity of the stream. As an illustration, Fig 8.3 represents a schematic drawing of a power plant located adjacent to a stream, with the power plant using a 205-acre cooling pond. Data about the stream at withdrawal point X are shown in Table 8.. These data are expressed as a frequency distribution relative to flow and concentration of total dissolved solids. The total dissolved solids are comprised primarily of chlorides. Water withdrawal from the stream into the cooling pond is 3.5 mgd. and cooling water usage in the power plant is 3.0 mgd, Water losses in the cooling pond occur from consumptive losses in the plant cooling cycle and from pond evaporation. Consumptive losses average 1.6 mgd, and evaporation losses vary from 0 to 1.4 mgd during

May through October. On this basis, total water losses in the cooling pond vary from 1.6 to 3.0 mgd. If the concentration factor in the pond is defined as "water in" divided by "water out", the highest concentration factor is 7 (3.5 mgd/0.5 mgd) and the lowest is 1.8 (3.5 mgd/1.9 mgd).

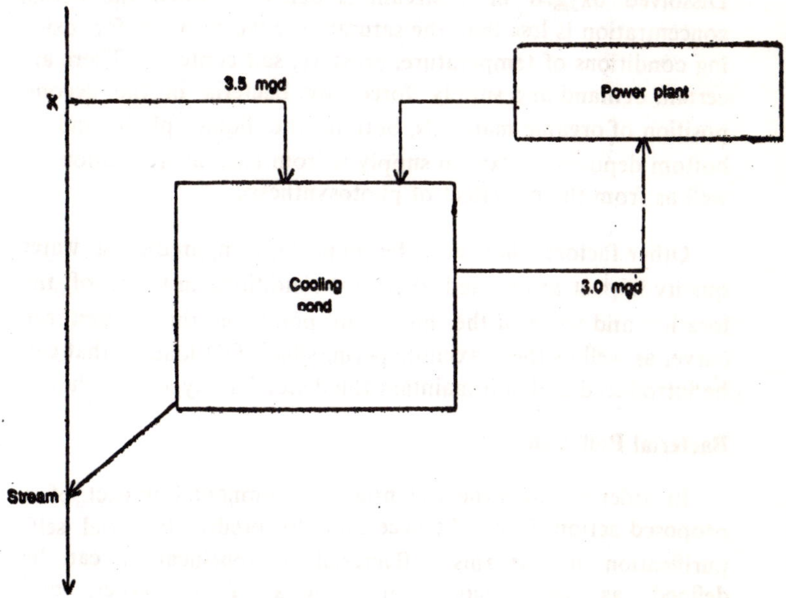

Fig. 8.2. Schematic drawing of power polant and water usage

TABLE 8.3

Summary of Total Dissolved Solids in Stream at Withdrawal Point

Stream flow	*Frequency*	*Total dissolved solids Mg/litre*
10.0	37	720
17.3	219	520
80.0	109	305

Nonconservative Pollutants

Nonconservative pollutants are organic materials, that can be biologically decomposed by bacteria in aqueous systems. One of the most common examples of stream pollution results from oxygen deficiency caused by organic waste discharges. Dissolved oxygen in a stream is deficient when the actual concentration is less than the saturation concentration for existing conditions of temperature, pressure, salt content. There are certain demand and supply forces by bacteria in the decomposition of organic materials, both in the liquid phase and in bottom deposits. Oxygen supply is from natural reaeration as well as from the net effect of photosynthesis.

Other factors that may be important in predicting water quality impact are related to critical conditions in terms of the location and value of the minimum point on the oxygen sag curve, as well as the maximum permissible BOD loading that can be introduced and still maintain the dissolved oxygen standard.

Bacterial Pollution

In order to assess the potential environmental impact of a proposed action, it may be necessary to predict bacterial self-purification in streams. Bacterial self-purification can be defined as the decrese of bacteria of all types, and especially those of fecal origin, as a function of flow distance or flow time in a river. Bacteria may starve, be devoured by predators, or be otherwise inactivated. Seasonal variation of coliform bacteria in municipal waste water are due to greater multiplications of intestional bacteria within sewage systems during summer months and also greater per capita discharge of these organisms in the summer months.

Thermal Pollution

Mathematical models of varying degrees of complexity have been developed to predict the persistence of heat in quiescent waters, flowing streams, estuaries, and the ocean. It is beyond the scope of this presentation to analyze the detailed aspects of each of these models.

STEP 11 : Pollution Control Measures

If it is determined that water quality standards are exceeded by the proposed action then abatement strategies or control measures should be presented. The details of various control technologies that can be used for minimizing pollutant emissions are described in another lesson.

STEP 12 : Discussion of Operational Impacts

This final step is included as a reminder to consider unique aspects of operational impacts relative to a variety of concerns. These concerns include the frequency distribution of decreased quality or quantity. fate of nutrients by incorporation into biomass, reconcentration of conservative pollutants into the food web, and chemical changes of certain inorganic chemical within aqueous systems. Chromium serves as an example of the latter since it is a transition metal that exhibits various oxidation states and behaviour patterns. Trivalent chromium is generally present as a cation, $Cr(OH)^{++}$, and is very chemically reactive, tending to sorb on suspended materials and subsequently settle from the liquid phase. Hexavalent chromium is anionc (CrO_4-) and chemically unreactive, thus tending to remain in solution. Changes can occur in the chromium oxidation state due to stream quality. For example, hexavalent chromium can be chemically reduced to trivalent chromium under anaerobic conditions, whereas chromium can be oxidized to hexavalent chromium under aerobic conditions.

9

Ganga Action Plan

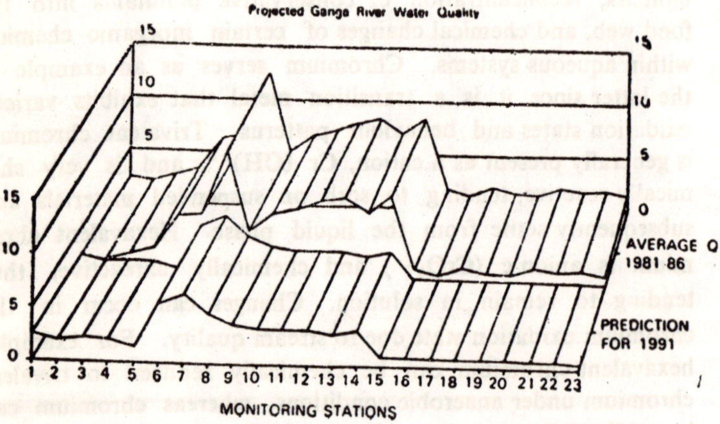

Introduction

Our rivers have been most precious to us since times immemorial. The alluvial soil brought by them has made their banks and adjoining plains so fertile that our country would never run short of food and fodder, if we use proper methods of cultivation. That is the reason why we reverse our rivers.

Pandit Jawaharlal Nehru once observed that he was looking for a man who could write the history of our rivers. Long after his death, turned up a professor who began to narrate his story of our rivers. Out of 14 major rivers, 12 are defiled with municipal and industrial wastes; the holy river Ganga is at best a sewage east of Sues.' Waves of concern rippled through the country. None could challenge the scientific data collected with utmost silence and meticulously colated. He worked out the conceptual remedy to restore Ganga to her health. It is now known as the 'Action Plan for the prevention of pollution of the Ganga'. It was initially Professor Nilay Chaudhri who thought of this.

The river Ganga, the life-line of millions of people has, over the years, been subjected to tremendous pressures. Most of its water in the upper reaches is diverted into canals; untreated sewage and industrial effluents are dumped into the river at numerous places and the residues of pesticides and insecticides used in the forms are washed into it.

This situation, already one of alarming propertions has been further aggravated by deforestation resulting in siliting, floods and reduced navigational possibilities.

Recognising the magnitude of this problem, and realising the importance of water quality as a cardinal element of management, the Government of India, in February, 1985 set up the Central Ganga Authority for the plannign and execution of a time bound programme to prevent the pollution of the river Ganga which is now better known as 'Ganga Action Plan'.

The basic principle for the plan is simple. The stoppage of discharge of sewave and sullage would reduce the pollution by almost 75 per cent. Class I cities (population over one lakh), 29 in number, cover 82.3 per cent of the total urban population living in urban settlements on the river banks The volume of sewage produced by them constitutes 88.5 per cent of the total volume of sewage flowing into the river. The construction of interceptors to divert the flow of sewage and other liquid wastes from Ganga is probably the surest way to reduce the pollution load. The proposal does not stop at that.

The dirty black sewage is simultaneously obnoxious and a resource. It is a very rich source of energy (through the production of biogas) and manuriol matter whose utilisation can increase the crop yield substantially. The action Plan envisages construction of treatment plants. 'Each treatment location should be viewed as a factory for resource recycling where sewage is the raw material, and energy in the form of biogas, manure, poultry feed, fish, and irrigant are marketable products. The items of resource recovery should not be an after throught but an integral part of the scheme.'

Ganga : The Holy River

No river evokes such feelings in the human mind as the Ganga does. She is loved not only by the Hidus but also by non-Hindus including the non-religious people for one reason or another. And all this just because she is 'Ganga'. Indeed, it is difficult to explain the charm she exudes, but the fact remains that the charm has remained intact through the centuries despite all the scientific progress that has taken place and the debunking religion has come in for. A British scientist was once shocked when he learnt of an Indian colleague taking a dip in the Ganga on an auspicious evening; he wondered how superstition had prevailed over science. What is even more interesting is the case of a non-Hindu, Russian scientist who plunged into the Ganga when someone pointed out the holy river to him. One wonders what the British socientist would have to say had he seen the sight. But then love is beyond reason.

The very name 'Ganga' conjures up a picture of holiness' of a peace and joy, of beauty and sweetness, of all that is uplifting to the mind of a Hindu. She is not just a river, she is a symbol of something higher than what the world has to offer. If Himalaya is a yogin steeped in meditatian, Ganga, rising from his heart is the culmination of that meditatian. She is a blessing for all, good or bad. Flowing through the plains the gives India both phy ical and spiritual nourishment. She is, in short, her life and soul. She is the bridge between heaven and earth. She is civinity in the form of water, according to Shri Ramakrishna, Industiies may polute her water but she remain as holy as ever, retaining her sancifying dowers.

According to Jawaharlal Nehru, the stay of Ganga is the story of India's civilisation and culture. It is indeed so, for India and Ganga both seem to be marching towards common goal the goal of truth. When she falls from the Himalayas, she is a tiny stream but when she meets the sea she becomes infinite. And it is by attoining this infinity that she attains liberatian, for there are no more banks to limit her size. Truth is another name for infinity. So it is truth that both India and Ganga are seeking.

But how does one attain truth ? By renouncing that which is finite, small and transient. As if to remind one of this, the waters of the Ganga have a yellow, gerua-like look, emphasising that truth is above everything else and that truth can be attained any through renunciation. Ganga is holy because the conveys this message.

Truth is also India's eternal message. God and truth are synanymous to a Hindu, the Ganga is dear to him because she is his mother, one who not only protects and reans him but also teachers and guides, it is difficult to imagine Hinduism without Ganga. According to Swami Vivekanand, the Gita and Ganga constitute, between themselves, the essence of Hinduism one its theory, and the other its practice.

GANGA : From Gomukh to Boy of Bengal

The Ganga rises in the snowbound heights of the Himalayas, nearly 4000 metres above sea level, from a dark. Icy cavern shaped like the mouth of a cow (known as Gomukh).

It protrudes out of a wait 100 m, high and drops sills of - ice and sheets of water. On the other side of the snout, is the Gangotri glacier which is more than 30 km, in length, 2 km, in breadth and 1000 feet in depth. How gigantic an ice-stream it is, so long, so wide and so deep ! It is said to be a remnant of the last ice age.

From Gomukh, the Bhagirathi flows on, now an imperuous river, now a cool crystal cascade, now a fuming, roaring torrent pushing on moraines, kicking off boulders and crashing ice till it joins the Alaknanda of Dev Prayag. From there onwards, the river calms down a bit and is called the Ganga.

The Alaknanda is also a very holy river, it rises from a glacier lying on the eostern face of the Chauknambha mountain. The Alaknanda is joined by a small river the Dhauilganga at Vishnu Prayag, by the Burhi Ganga at Guptaprayag, by the Nandakini at Nandprayag by the Pincda at Karanprayag and by the Manciakind at Rudraprayag. All these prayagas are considered holy by the hindns ; so is Devprayag. There are six prayagas in the Himalayas. The seventh once is at Allahabad where the Ganga is joined by The Yamuna. After Allahabad, down the stream there are no prayagas though many rivers fall into the Ganga.

The Ganga and its tributories have a peculiar tendency to flow for some distance parallel to the Himalayan ranges if deep ravines but then they take acute band and flow in deep gorges transversely. These gorges are sometimes hundreds or metres deep and appear owesome from above, and the river below looks like a thread. One other feature of these rivers which may be interesting to you is that the Bhagirathi flows initially from east to west and then takes a turn from west to east, whereas Alaknanda flows initially from west to east and then turns, and flows east to west till it joins the Bhagirathi at Devprayag. Thus, these two rivers flowing in opposite directions join to form a garland at Devprayag.

At Devaprayag, the Bhagirathi is joined by the Alakananda (which originates above Badrinath near the Indro-Tibet border and receives the waters of Mandakini at Rudraprayag) to form the great Ganga river.

From a swift moving stream the Ganga grows into a wide river just before it touches the plains at Rishikesh. Then begins its over 2000 km-long journey through the heart of India the Indo-Gangetic plain—before it falls into the Bay of Bengal.

The Ganga enters plains at Hardwar which is called 'Haridwar' by the Vaishnavites and 'Hardwar' by the Shaivites, for the word 'Hari' means Vihnu and 'Har' Shiva, and 'Dwar' means gate. From Hardwar, the Ganga starts slowing down its pace. Its speed is reduced substantially at Kanpur and Allahabad, and is further reduced at Patna and still further at Monghyr. After that, the Ganga flows very sluggishly till it falls into the Bay of Bengal.

On its way, the Ganga is joined by many rivers: by the Ram Ganga at Kannauj, and by the Yamuna and the mythical, invisible Saraswati at Allahabad.

After Allahabad, the Ganga receives several major tributaries at more frequent intervals: the Tons and Sone, the Gomti, the Ghagra, the Gandak, the Burhi Gandak and the Kosi.

Further east, beyond Farakka in Bangladesh, the Ganga branches into the Padma. Here, once again, Ganga takes the name Bhagirathi and winds down the last stretch-the delta of West Bengal, before it merges with the Bay of Bengal under the name Hoogly.

In its 2525 km long course, the Ganga dramatically varies her pace and form. No more than a few feet wide in the Himalayas, it is a tumbling, boisterous stream in the hills, and a deep, fast flowing river at Rishikesh where it enters the plains. In the plains of Uttar Pradesh, it is a bed of sand in summer. In Bihar, it is a vast expanse of water, 4 to 14 kilometers in width.

The tributaries of the Ganga account for 60% of the total water of the river. Though the Ganga and the Yamuna are often reffered to as the twin sisters, in terms of flow, Ghagra is. the more important tributary—20% of the Ganga waters are fed by Ghagra as against 16% by the Yamuna.

Ganga : Its Importance

No other river affects the lives of so many people in so many states of India as the Ganga. It is the largest river basin in our country covering over one-forth of its total surface area. From the source to the sea via the Bhagirathi-Hooghly it is a run of over 2,500 km.

The Ganga is among the first 10 mighty rivers in the world in average annual run-off measured in terms of thousand million cubic metres. Its basin has a total population of nearly 45 million, of which about 50 per cent live in Uttar Pradesh, ten per cent in Bihar and 40 per cent in West Bengal.

It is obvious why the fate of such a river should be equated with the destiny of the millions who live on her banks or its numerous tributaries; why fluctuations in its flow are watched over with so much anxiety, its pollution creates near-panic and share of its water among contending claimants generates so much of heat.

Strictly speaking, there are eight states in India which form parts of the Ganga basin, or more precisely, together act as the catchment area of the main river. But the recipiens of major benefits are, by and large, only the three states of Uttar Pradesh, Bihar and West Bengal.

Among all the river basins in India, the Ganga basin has the largest irrigation facilities in which canals play a dominant role. There are four major irrigation projects in operation, all in U.P. *viz.*, the Upper Ganga Canal System, Lower Ganga Canal System, the three pumped canal System, the three pumped canal schemes of Dalmau, Bhoupali and Zamania and the Ramganga project, to increase the irrigation capacity of the Upper and Lower Canal Systems. To augment further the role of the Upper and Lower Canal Systems, the Tehri Dam and the

Kotli Bhel Dam are being constructed. Moreover, a new canal system is now being introduced in Uttar Pradesh known the Madhya Ganga Canal System. Thus, it is obvious that the entire irrigation system of U.P. hinges on the supply of water from the Ganga.

If the Ganga is the mainstay of irrigation in U.P. it has been a supplier of precious fresh water to industries in all the three riparian states. According to a survey conducted by the Central Water and Power Commission, the hydro-power generation potential of Ganga lies mainly in the U.P. Himalayas and this has been largelly neglected area. However, four major thermal power plants in the country depend on Ganga water—at Pankhi in Kanpur, Barauni in Bihar, Farakka in Murshidabad and Bendel in Hooghly.

Uttar Pradesh is a growing industrial state and is favoured by large investments from the Centre. Kanpur is well-known for its cotton, jute and woollen textiles, Varanasi for silk and light engineering Allahabad for light engineering while hardwar has made its mark on the industrial map of India for the production of antibiotics and heavy electricals. Moreover, U.P. is one of the principal sugar producing states in the country with factories dotting the banks of the Ganga throughout its 1,500 kilometre run in the state.

Comparatively speaking, Bihar is neither a substantial user of Ganga water for irrigation nor for industries. In its 450 kilometres stretch through Bihar, only in Patna and Bhagalpur are there are some uses by small and light engineering industries.

West Bengal's industrial economy, on the other hand, has been largely dependent on Ganga water since the founding of Calcutta nearly 300 years ago. For the Calcutta Metropolitan District, the hub of industries in the state, the prime supplier of water is the Ganga. Decrepit but still the most important agro-jute is concentrated on the banks of the Bhagirathi Hooghly together with a large number of light and medium engineering factories.

It is generally not realised that though not comparable to the needs of agriculture, industry also requires a significant amount of fresh water. By 1980 the total requirement of such water in the country was about 12,000-15,000 million cubic metre and at the end of this century this amount is expected to go to 35,000 million cubic metres. The shar of the three riparian Ganga states will be considerable, particularly of U.P. and Bihar, because the CMD has almost reached satuaration point in industrial congestion.

The third important area of the Ganga's contribution to the north Indian economy is supplying water for domestic and municipal requirements. One would do well to remember that the Ganga plain has the highest concentration of people in India. It has 29 class cities, 71 class II and class III towns and numerous small towns belonging to Census categories of IV, V and VI. But more numerous are its rural population.

It is assumed that domestic requirements of water for urban and rural populations in a tropical country like India are 60 gallons per day per capita for the former and half of that for the latter. It has also been calculated that by the end of this century the demand for domestic water from the Ganga will rise to about 62,000 million cubic metres. U.P. once again, will be the major claimant for the supply of such water.

Lastly, there is the demand for navigation, particularly for the existence of the Calcutta Haldia Port complex. The quantum agreed upon by all experts is a minimum flow of 45,000 cusecs all the year round through the river at the Calcutta port site. Add to this the needs of fish, wildlife and for recreation.

If we calculate the existing share of the different demands on Ganga water, it becomes clear that of present the largest claimant (over 75 per cent) is irrigation, the rest being shared among industries, domestic and municipal demands, power and transportation. It is obvious that this is on unbalanced situation and that with such an overwhelming demand for Ganga water by agriculture, particularly in the upper reaches of the river, the crisis of the Calcutta-Haldia port complex is bound to persist and deepen. Such is the role of the Ganga in India's life.

The alarm sounded about Ganga water pollution has lately added to the complexity of the present situation. To maintain and strengthen the vital role played by the Ganga in the life and economy of the country, a four-pronged approach has been advocated by experts. The first among these concerns the problem of augmentation of flow in the river, which is the most vital single step for its rejuvenation. The second aims at pollution control. The third and the fourth deal with the controlled use of water not only territorially, but all sectorally.

It is known that despite its being a flow resource of considerable magnitude, the increasing demand of the growing population in its basin makes this resource increasingly more scarce in the lean months when the volume of flow is substantially reduced. With the onset of the rainy season and extedded over to a few months afterwards the flow is much more than we shall be able to consume even with higher demands in future. Most of it flows into the sea.

The problem is to prevent this waste; to use the excess water to augment the flow in the dry period, thus ensuring a continuous and steady supyly of Ganga water throughout the year.

This involves detailed planning for the storage of excess water by a string of dams all along the Himalayan foothills in Uttar Pradesh and Bihar and possibly a linkage between the Brahmaputra and the Ganga. There is already a proposal of a system of interlinking canals to pour the surplus flow of the Kosi, Gandak, Burhi Gandak and the Ghagra to the west into the Ganga, construction of a number of dams in Nepal Tehri and at other places.

The scheme is time consuming and entails heavy expenditure. But it can be phased out in the coming plan periods as we have done in many cases.

Two other aspects of the plan for the Ganga should be to introduce rational use of water sectorally as well as teritorially. In all advanced countries, river water is budgeted carefully and allocations made for the four major users-agriculture, industry, domestic and municipal requirement sand transport. Runaway use by one sector at the expense of all others is not allowed, for in the ultimate cost-benefit accountancy, it is not only wasteful but a'so extremely dangerous and harmful for the country in the long run.

Same is the problem with territorial distribution of possible uses. Even leaving out the knotty question of whether the Ganga is an international river and whether there should be proper sharing of its water between different countries, the fact remains that the Ganga certainly is an interstate river and that all users have the same claim over its water. Such claims cannot be wished away as narrow, regional or parochial. There is as much justification for U.P. farmers to draw irrigation water from the Ganga as the people of Bihar asking a much larger share of this water for agriculture and industry or the people of West Bengal petitioning for 40,000 cusecs all the year round to save the Calcutta-Haldia port complex.

Unfortunately for us, Indian planning so far has been largely one sided and partial depending on the strength of the lobby which can move the planning mechanism of the Union government most effectively. Take for example the case of irrigation projects

One very competent and high-placed officer in the Union ministry of agriculture wrote that our irrigation plans were drafted in isolation and failed to take note of the many other vital questions linked with the construction of dams and canals for the use of water. As for example, our irrigation projects plan for the supply of water but not drainage, construct dams but do not envisage maintenance of water sheds, dig the main canals but do not plan detailed field distribution, avoid making conjunctive use of surface and ground water in order at conserve our water resources to the maximum possible extent, supply water to one region and forget about the others.

So far the line has been to use short cuts without a comprehensive understanding of the problem. The result may prove to be disastrous in the coming years of which there are ample indications already. It is, therefore, of utmost importance to realise that a proper regional and sectoral budget for the use of Ganga water is in the interest of the nation as a whole.

Lastly, comes the question of pollution. So much has been written on Ganga water pollution in recent months that we need not elaborate on the problem here. It can be noted th t though pollution of Ganga water is a big problem by itself, it cannot be tackled seriously without augmenting the flow or maintaining a strict control on the use of its waters by sectors or regions.

In short, there is a strong case of advocating a policy of planned use, conservation and development of Ganga water. Not only to meet the problems of today but more so of the future. A piecemeal approach, which has become the practice with our policy-makers, may irreparably damage this supremely important flow resource of India and deliver a crippling blow to the health and economy of one of the largest and most densely settled river valleys in the world.

Ganga : Overuse, Abuse and Pollution

'Ganga Tera Pani Amrit'. Since the dawn of history, the waters of the Ganga have fed the crops, washed the dirt, quenched the peoples' thirst, and have been considered to be nectar. This belief, that the Ganga is 'Amrit' or nectar has not, however, prevented the holy river's overuse, abuse and pollution.

This situation is further compound by a basic meteorological fact. Three-fourths of the annual rainfall in the Ganga basin in the 4 months from June to September. In the remaining 8 months, most of the basin is dry-the time when the fields have to watered.

When Cautley in 1843, started digging the upper Ganga canal from Hardwar, in order to find a solution to the above problem, people feared he was playing with sacred waters. The canal, today, has made all the difference to the doab between the Ganga and the Yamuna, which was previously affected by famine. At the same time, however, it has taken away more than 90% of the water of the Ganga during the lean season. What remains in the river between Hardwar and Prayag is merely a trickle. Understandably then, the need of the hour is not only to check water pollution, but also to take measures to conserve water itself.

What is River Pollution ?

In simple terms pollution makes the water unsuitable for a specific use. When the water is very clear as in Devaprayag, we can drink it straight from the river. When it is a little muddy and mixed with dirt and waste, we can neither bathe in it nor drink it.

What are the Principal Sources of Pollution ?

– The waste waters from our towns and villages, the drains and sewers that carry these wastes to the river. It is estimated that 900 million litres of sewage is dumped into the Ganga every day.

– The garbage we throw by the riverside or sometimes even into the river

– The chemical and water wastes from industries.

– The harmful residues or insecticides and pesticides draining away from the fields.

– Throwing of animal carcasses and half-burnt and unburnt human carcasses into the river.

– Defecation on the banks of the river.

The causes are almost endless, but all these acts of pollution are not accidental. We may not deliberately wish to pollute the Ganga but somehow end up doing so.

How to Measure the Extent of Pollution ?

Scientists usually measure river quality with three standards. One is Dissolved Oxygen (DO); the more the oxygen, the better the quality. The second is Biochemical Oxygen Demand (BOD). This is a measure of the degree of pollution in the river. If BOD is high, the quality of river water is low. The less the BOD, the better.

The third yardstick is the probable number of coliform in the water. Coliform are bacteria. All of them do not cause disease but faecal coliform and pathogens usually do.

These pathogens can cause cholera, jaundice, typhoid and various other intestinal diseases. Because of powerful drugs and organised medical care, cholera does not usually become an epidemic but in most of the Gangetic belt water-borne diseases are widely prevalent.

Pollution, however, is not the only problem faced by the river Ganga. Silting is an equally serious one and deserves urgent remedial action.

The watershed of Ganga in the ancient days was known for its rich forest cover. But deforestation has been rampant in the past few decades. As a result, rich top soil is carried away every year from the hills to the river and then to the sea. It is estimated that three tons of silt is carried away from every square kilometer of the Ganga's catchment areas.

Silt deposit in the river raises the river bed, causes floods and in the lean season the flow becomes stagnant. Floods in the monsoon, no water in the dry season. This has become the story of the Ganga in many parts of the country.

Concern for the River

Concern for the environment may have been a part of our ancient heritage, but we have forgotten much of that in recent years. Our late Prime Minister Indira Gandhi re-established this concern in the public conscience. Time and again she told the people the simple but basic truth that the Survival of human life depended on the survival of plant and animal life.

Brought up on the banks of the Ganga, she shared with her father an attachment to this great river. In 1980 she asked the Central Pollution Control Board to make a study of the source and extent of pollution in the Ganga and to come up with a programme for its prevention.

The study of water polluion in the Ganga basin prepared by the Central Pollution Control Board in 1984 is the Country's first comprehensive assessment of the problem. The report reveals :

—Three fourths of the pollution of the Ganga is from untreated municipal sewage.

– Out of this 88% comes from 27 Class-I cities.

—Only a few town have sewage treatment facilities but even these are not functioning.

—Industries accounted for 25% of the total pollution, but this was far more acute in areas such as Calcutta and Kanpur.

Formulation of the Ganga Action Plan

In any river quality management programme, the first step is to identify the sources of pollution. The Central Pollution Control Board identified such sources through its exhaustive study of the river basin published under the River Basin Series. This is the first systematic document which formed the basis for the Ganga Action Plan. Fertilizer and pesticide use, industrial pollutional load to the river, land use pattern, hydrological aspects and river classification are detailed in the document.

This inventory of pollution formed the basis for the Ganga Action Plan document prepared by the Department of Environment in December, 1984. Since the inventory had identified domestic wastes from the 27 Class-I cities along the river Ganga as the major cause of pollution, the Action Plan's thrust is to divert such wastes away from the river. The treatment of waste waters has been envisaged as a resource recorvery mechanism. Bio gas as a principal source of energy, the application of treated effluents as an irrigant and a source of pisciculture and aquaculture were envisaged as salient features of the Action Plan.

Addressing the nation on 5th January, 1985, the former Prime Minister Rajiv Gandhi declared, "The Ganga is the symbol of India, the source of our legend and poetry, the sustainer of millions. Today it is one of the most polluted rivers. We will rerestore the pristine purity of the Ganga".

Within one month of the announcement, the Central Ganga Authority was constituted. In June, 1985 a separate wing of the Department of Environment was created to administer the programme. A sum of Rs 240 crores has been earmarked for this purpose under the Seventh Plan.

Few countries in the world have taken up a project a challenging and daunting as this. The Prime Minister's declaration has been received with great enthusiasm by the people. Interest in the project is strong, both at home and abroad.

What does the Action Plan Seek to do ?

Very simply put, it proposes to divert the sewage now flowing into the river to other location for treatment, and convert it into a valuable energy source. For example, treated sewage for use in fish ponds and aquaculture, irrigation and in generating bio electricity.

The various schemes proposed under the Action Plan have been planned to reduce the pollution in the Ganga by at least 75%.

The Ganga Action Plan provides the first opportunity in years for the country to have an integrated view of this premier river system.

Programme of Works

The Ganga Action Plan comparises 262 schemes at a cost of about 260 crores. Their estimeted cost are indicated below :

(Rs. in crores)

	No.	*Cost*
Interception and diversion of waste water	90	99.50
Sewage Treatment Plants	32	105.08
Low Cost Sanitation	42	20.40
Electric Crematoria	29	10.[illegible]4
River Front Facilities and related improvements	34	13.57
Other schemes for biological conservation of aquatic species, river quality monitoring, etc.	35	10.37
Total	262	259.27

Towns Covered ender the Ganga Action Plan

The towns to be covered under the Ganga Action Plan are as under :

U.P.	*Bihar*	*West Bengal*
1. Hardwar-Rishikesh	Patna	Berhampure
2. Farukhabad-Fatehgarh	Chhapra	Nabadwip
3. Kanpur	Monghyr	Calcutta, Howrah and other
4. Allahabad	Bhagalpur	Class-I towns of the Calcutta
5. Mirzapur		Metropolitan District
6. Varanasi		

Kanpur

Kanpur, one of India's earlist industrialised towns, is an important centre of textiles, engineering industries and tanneries. During the early days of industrialisation, Ganga became the source of transportation as well as the sewer. Today Kanpur is the second largest city along the Ganga and is an important centre for higher technical, education, but its infrastructure continues to be primitive The main city and its adjoining areas are partly sewered, but many sewers are choked rendering the system inoperative. Sullage and sewage is carried into 13 storm water nallahs for final disposal into the river. The main sewage pumping plant at Jajmau is worn out. The 70 odd tanneries in and around Jajmau mix domestic and highly toxic tannery/waste and carry it to the river. In all of U.P., the stretch of Ganga at Kanpur is the most severly polluted.

Under the Ganga Action Plan, the pumping station at Jajmau is being renovated and its capacity expanded. Improvements to the sewage farm have been taken up. A major MLD treatment plant is to be built. An integrated sanitation scheme for Jajmau tannery area to treat the tannery wastes and improve drainage and garbage disposal has been initiated. A phased programme for tapping waste waters from the nallahs and the cleaning/renovation of sewers has commenced.

Allahabad

Allahabad or Prayag is where the Ganga, the Yamuna and the mythical Ṣaraswati converage into one. Domestic waste rather than industrial effluent is the problems here. The Gaughat pumping station is being renovated. A sewage treatment plant at Naini is being constructed. Extensive renovation of the sewer network has also commenced.

Varanasi

One of the oldest, continuously inhabited cities of the world, Varanasi is regarded as a leading centre of religon and philosophy, music and dance, arts and crafts and of higher learning.

Unfortunately, it is also a leading source of pollution as well, accounting roughly for 2 % of UP's total pollution load on the Ganga.

Varanasi's 400 km long sewerage system is over 75 years old and most of its underground drains were built 200 to 300 years ago. This ageing sewerage and drainage system has virtually collapsed with the rapid growth of the city and its population.

The Integrated City Development Programme for Uaranasi seeks to address the problems of the river, the river from and the city which are indistingusihable in daily life. The old treatment plant at Banaras Hindu University has been made functional and its capacity is being expanded; another at Diesel Locomotive Works is under construction and the third at Konia-Dinapur is being taken up for construction. The sewage pumping stations at the ghats have been renovated. Extensive renovation of the sewer networks is under way. Public facilities at several of the ghats are being improved including the provision of Sluabh Sauchalayas. The defunct kunds and tables are being re-excavated. A major improvement of the garbage collection and disposal system has been taken up. Special arrangements are being made to deal with the problem of throwing of dead bodies and animal carcasses.

For the first time in recent years, a determined, multi-sectotal approach has been initiated to improve conditions in this historical city.

Patna

Emerging from another great the ancient city, Pataliputra, Patna today is the capital of Bihar with a population close to a million The main source of pollution is the city's sewage. About 50 million litres per day flows into the river. One of the largest sewage treatment plants was built at saidput-Patna. But it became defunct more than 10 years ago. Under the Ganga Action Plan, its renovation was given high priority. The paint was recommissioned in June, 1986. Another plant at Beur has been similarly brought back into use. Supported by funds from the Ganga Action Plan, the State Water & Sewage Board has taken up the renovation and construction of pump houses and main sewers in different parts of the city. A major river from scheme is also to be taken up.

Calcutta

In much of metropolitan Calcutta, growth and pollution have almost been synonymous. Apart from Calcutta and Howrah, connected by the famous steel bridge, there are 37 towns and nearly 150 factories, big and small, all of which use the Ganga (known as Hoogly here) as their drain. The wastes from the factories alone is estimated to cuse a pollution load of 37 tons a day. The CMDA and other agencies of the West Bengal Government have planned a number of schemes to deal with the problems and commenced work on many. At Titagarh, Bhatpara, Chandannagar and Serampore existing treatment plants and sewer systems are being renovated. New sewage facilities including treatment plants are planned for Calcutta city and adjoining areas. Low Cost Sanitation schemes have commenced in many towns. Construction of crematoria in important locations and river front development schemes have also been taken up under the Action Plan.

Special Features of the Ganga Action Plan Sewage Treatment Plants

Under the Ganga Action Plan, a total of 10 sewage treatment plants are to be renovated and another 21 new plants are to be constructed in different locations While many of the STPs are based on established technology such as activated sludge, trickling filter etc., in few locations alternate technologies are being applied. For example up flow An aerobics Sludge Blanket treatment technology will be used at Jajmau, Kanpur and at Mizapur as part of the Indo-Dutch Integral Sanitations Project. The aerated lagoon proces is being used in Bhagalpur. The Rotating Biological Rope Contractor process developed by NEERI, Nagpur, for small communities is being used in one of the treatment units at Rishikesh. Large capacity archimedean sorew pumps are to be used in the Konia main pumping station at Varanasi.

Bio-energy is an important component at sewage treatment plants under the Ganga Action Plan. Such components have been incorporated in the design and tenders for the treatment plants in 5 locations already. Provision for bio-energy has been kept in 6 more locations. Apart from bio energy the use of treated effluents as an irrigant and the facilities for this purpose are being improved and expanded in 24 locations. Development of pisciculture with treated effiuent is also being specificaly provided in 8 locations.

River Front Facilities

As part of the Action Plan, schemes have been taken up at certain locations for the development of the river front areas for preventing the pollution of the river. In Varanasi the schemes relate to river front development, protection of slopes of ghats, construction of community toilets, improvement of lanes and by-lanes, provision of street lighting, etc. In Allahabad, a major project for the integrated development of the river front has been initiated. Similar schemes are to be taken up in certain other locations also.

Low Cost Sanitation

Since in many towns the sewerage network is not extensive considerable outlay has been proposed on low cost sanitation facilities involving construction of public toilet complexes, construction of new pour flush latrines and conversion of dry latrines in almost all the locations taken up under the Ganga Action Plan. Cost of schemes identified and sanctioned so far is Rs. 27.43 crores. It involves construction of 3,026 public toilet complexes and conversion/construction of more than one lakh new pour flush latrines.

Construction of Electric Cremataria

Lrowing of unburnt or partly burnt dead bodies and animal carcasses into the river is considered to be a cause of pollution in specific locations. To deal with the problem, UP the State Government has sanctioned construction of an electric crematorium at Varanasi. Under the Ganga Action Plan a total of 29 cremataria has been sanctioned for different locations.

Apart from improving the sanitary conditions near the river at these locations, it is expected that the use of electric crematoria will be cost effective and save considerable quantity of fire wood now used for this purpose. The local bodies will be responsible for the operation and maintenance of the these facilities.

A scheme for disposal of animal carcasses of Varanasi has also been identified as part of the city development package under the Ganga Action Plan. Besides, the scheme for improvement of facilities for the river traffic police is also expected to go a long way in preventing dumping of animal carcasses, and other abuses of the river.

Co-Financing & Pooling of Resources

While the cost of certain schemes connected with the prevention of the pollution of the river Ganga is being shared with other concerned agencies, resources have been pooled from different sources to create permanent facilities in certain locations in the context of implementation of the Ganga Action Plan. The cost of construction of the sewage treatment plant at DLW, Varanasi is belong shared with the Railway Ministry, Similarly, the cost of costruction of a new road bridge across Yamuna is being shared with the Ministry of Surface Transport. In the context of the Jawaharlal Nehru Birth . Centenary Celebrations, a special scheme to render the whole town of Allahabad scavenger free, (on idea which had been mooted by Pandit Jawaharlal Nehru himself when he was the Chairman of the Allahabad Municipality in 1922-23) has been taken up. The Ministry of Welfare, U.P. Government, the Ganga Action Plan and the Allahabad Nagar Mahapalika will contribute Rs. 7 crores to this scheme. Another major project which has been initiated is for the integrated development of the river front between Baluaghat and the historic Akbar Fort at Allahabad. The scheme which has been prepared with the help of the School of Planning and Architecture, Delhi will be financed through Ministry of Tourism, U.P. State Government, and Ganga Action Plan.

Pollution from Industries

While the Action Plan has focussed on pollution from municipal wastes, the Central Ganga Authority has also directed that special steps should be taken to deal with serious industrial pollution in specific locations. The steering Committee under the Action Plan had set up an Expert Group which reviewed the position of 264 industrial units located in the river banks. The group identified 68 units as gross polluters considering the volume of discharge (more than 1 MLD) and the characteristics (high content of toxic materials). These 68 units have been asked to set up treatment facilities in a timebound manner. Of these, 34 are in U.P. 5 in Bihar and 29 in West Bengal.

Out of these 68, when the GPD commenced work, only 15 had provided effluent treatment facilities. The position obtaining now is as under :

(i) Treatment plants provided	25
(ii) Treatment plants under construction	11
(iii) Units closed down	5
Total	41

Action against the remaining 27 units have been initiated to ensure the setting of treatment facilities.

Scientific Content in the Context of Ganga Action Plant

Several steps have been taken to secure the inputs of the Indian scientists and institutions in the execution and monitoring of schemes under Ganga Action Plan. The focus of the scientific content of Ganga Action Plan is on the following :

Promotion of Research in Relevant Areas

A Research Committee for the Ganga Action Plan has been constituted to aid and advise the Ganga Project Directorate in the matter of promotion of research and technology application to further development of holistic policies and practices for conservation of the resources of the river. The State Government have constituted State level committees of scientists to secure interaction of scientists in the formulation of schemes under the Ganga Action Plan.

Interaction with Scientists and Technologists

In collaboration with the leading national laboratories and with the assistance from the leading scientists several workshops have been held in areas relevant to Ganga Action Plan. For example :

* Land treatment of sewage and application of sludge.
  — IARI, New Delhi.
* Recycling and reuse of waste water in industries.
  — CEI, New Delhi.
* Biological conservation and biomonitoring in receiving waters.
  — Jawaharlal Nehru University, New Delhi.
* Water Quality Modelling and treatment technologies.
  — IIT, Bombay.
* River front conservation and instream uses.
  — School of Planning & Architecture, New Delhi.

Organisational Arrangements and Monitoring of Progress

Central Ganga Authority, under the Chairmanship of Prime Minister has been set up to determine policies, allocate funds and oversee progress. A steering Committee is responsible for approval of schemes and annual allocations and reviwe of progress regularly. Another Committee under the Chairmanship of the Member (Science), Planning Commission monitors the progress of schemes and reports to the CGA.

The Ganga Project Directorate functions as a wing of the Ministry of Environment and Forests to appraise and sanction schemes and to coordinate the implementation of the Action Plan in the three States of U.P. Bihar and West Bengal.

In each of the three State Governments, a nodal department is responsible for coordinating the preparation and execution of schemes. In U.P. and Bihar, it is the Department of Urban Development and in West Bengal, it is the Department of Environment. Agencies such as U.P. Jal Nigam, Bihar State Water Supply and Sewage Board, Calcutta Metropolitian Development Authority, are some of the agencies executing this work.

Regional offices of the Ganga Project Directorate have been set up in Allahabad and Calcutta to facilitate liaison with the three State Governments and to monitor the progress of work.

The State Governments have also set up a State Level Task Force under the chairmanship of Chief Secretary to periodically review the progress. City Task Forces and local project managers coordinate work at the local level.

Monitoring of Progress

A detailed format for monitoring the progress has been evoived in consultation with the implementing agencies. Each scheme is monitored with respect to its monthly target of physical and financial progress so that bottlenecks are identified and remedial measures taken. For major locations such as Hardwar Rishikesh, Kanpur, Allahabad, Varanasi, Patna, Hawrah and Calcutta Municipal Corporation, citywise consolidated progress charts and schedules have been prepared.

Monitoring of Water Quality

River water quality in India is assessed on the basis of standards fixed for different uses. The Central Pollution Control Board has fixed such standards for different uses, sush as Class A (fit for drinking water supply with chlorination), Class B (fit for bathing), Class C (fit for drinking water supply after full treatment), Class D (fit for aquatic life) and Class E (fit for receiving wastes) etc.

The study conducted by central Pollution Control Board indicated that of present the water quality in Ganga in many parts below Hardwar was Class C or D. The Action Plan objective is to bring river quality to Class B. Regular monitoring of river qualify is essential not only to assess the problems faced but also to measure the impact of the Action Plan objective of water quality.

The arrangements for regular monitoring have been strengthened substantially. With assistance from GPD, the following agencies are engaged in the work :

Agency	*Responsibility*
Central Board for Prevention and Control of Water Pollution, New Delhi.	Sampling, collection and analysis at 27 identified locations Data storage and preparation of reports to Monitoring Committee. Quality control in respect of analytical parameters.
Industrial Toxicology Research Centre, Lucknow.	Analysis and reporting of 10 heavy metals and 7 pesticides parameters.
U.P. State Pollution Control Board, Lucknow.	Intensive micro-level monitoring at Hardwar-Rishikesh, Allahabad, Kanpur and Varanasi.
Bhabha Atomic Research Centre, Bombay.	Analytical qualify control in respect of heavy metals.

Public Involvement in the Ganga Action Plan

While launching the Ganga Action Plan at Varanasi on the 14th June, 1986 our former Prime Minister had stated very clearly. "The Ganga Action Plan is not just a Government plan. It has not been prepared for the PWD or government officials alone. It is a plan for all the people of India, one in which they can come forward and participate. It is upto us to clean the whole of the Ganga and refrain from poluting it".

In the first and the last analysis Ganga will remain clean only if the people want if and help in the process. There are many ways in which this can be done.

We must know what are the problems of river pollution, what are its sources and why pollution should concern us. We must know how pollution affects the use of the river water, how if can affect our health, how it damages fish and other aquatic life.

In our daily use of the river, we must avoid doing certain things.

For example :

Dontt delecate on the banks of the river.

Don't throw garbage into the river.

Don't use shampoos and detergents in the river.

Avoid washing cattle in the river water. Don't wash clothes in the river.

In promoting public involvement in the programme, the man objectives have been the following.

(a) Promote awareness of the problems of pollution of rivers in general and Ganga in particular;

(b) Disseminate the importance of the initative taken by the Prime Minister towards cleaner Ganga;

(c) Promotion of right attitudes towards use of rivers;

(d) Improve motivation and performance on the part of those engaged in the Ganga Action Plan.

(e) Promote awareness of erosion related issues such as the slitation of rivers, soil erosion, afforestation, industrial pollution, conservation of water resources, etc., and

(f) Secure public cooperation and participation in the implementation of the Action Plan.

The involvement of the people is being sought through public information and awareness campaigns utilising multiple media, obtaining public cooperation in specific schemes and organising direct public participation in some types of work taken under the Ganga Action Plan.

School Children can do a lot to spread this awareness. There are special programme to help secondary schools along the Ganga to organise simple water quality monitoring tests. We have specially prepared kits and manuals for this purpose. Already about 90 schools and 20,000 school children have participated in this programme.

Other programmes for school children such as debate, essay, painting and song and dance competitions have been organised in Allahabad, Patna, Varanasi, Hardwar and other places.

Youth and Ganga is another special feature of the awareness programme, NSS, NCC, Nehru Yuvak Kendra and other youth volunteers have held Ganga Sewa Shivirs in several locations so far. Over 4000 young people have helped in free planting cleaning of the ghats, shramdam for river front development works and so on.

Males and festivals are very much a part of life along the Ganga. Special exhibitions on the Ganga Action Plan have been organised to coincide with congreations of pilgrims in Hardwar, Allahabad, Mirzapur, Varanasi, Sonepur, Sultanganj, Naihati, Serampur and Ganga Sagar. Special cultural programmes addressed to the pilgrims relating to the theme of environment and prepared in folk media have been held through North Zone Cultural Centre and the INTACH.

Talk to your neighbours, your children. your friends. Youth Clubs, Sports, Organisations Welfare Associations, Women's Groups, pilgrims - all can help to organise programmes of public awarness, public cooperation and public participation.

Recycling Sewage : (Where we went wrong).

The Action Plan published by the department of environment (government of India) is accompanied with a diagram of the scheme which shows the sun shining at the top in all its glory. Who can ignore the sun in India ? It is also a source of nonpoluting energy and light. It should also be utilised. The light energy received on a horizontal surface is expressed in an unit known as 'Langely', named offer the American astronomer and aeroplane pioneer Samuel Pierpont Langley (1834-1906). The Langley unit gives the amount of heat (calorie) received by an unit surface area (square centimetre) in an unit time (day), India, situated between 8° and 33° north latitudes, receives between 200 to 650 Langley units per day. The Nagpur-Kanpur region receives 320 to 480 such units. In dusty urban locations 6 to 10 per cent solar energy gets absorbed in particles floating in the air. Even then, the heat is not small. If the total Ganga water surface in the plains, say 2,000 km, is taken, an astronomical amount of heat is received by the river. The immediate effect is the rise in water temperature.

A Ganga Factsheet

Length of the Ganga; 2,525 km.

The Source of the river is at Gangotri, Uttar Pradesh, at an elevation of 7.010 metres.

The flow from Gangotri is regarded as the main flow of the Ganga. At Devaprayag it meets the combined flow of the Alakananda and Mandakini to become the Ganga as we know it futher downstream.

The are 100 towns located on the Ganga : 26 in Uttar Pradesh 15 in Bihar, 59 in West Bengal. Of these 100 towns, 29 have a population of more than one lakh each and by virtue of this are classified as class I cities. There are 23 class II cities with a population of between 50,000 and 1,00,000 each and the remaining 48 urban concentrations have a population of 58,000 or less each.

It is estimated that 43 per cent of the total irrigated area in country is located in the Ganga basin.

Much of the dry weather flow of the river is diverted to the Upper Ganga Canal at Hardwar and again into the Lower Ganga Canal at Aligarh. Both canals were built for irrigation and were completed in 1854 and 1879 respectively.

These two canals take away a great deal of the dry season flow, leaving the river vulnerable to pollution. But lower down, the Ganga receives 60 per cent of its water from the Yamuna, Ghaghara, Kosi and Gandak, which meet the river at or below Allahabad.

Though the Ganga is given so much importance in India, it does not rate an entry among the great rivers of the world. The Nile, Amazon and the Misissipi are two and a half time longer. The Congo Yellow River, Mackenzie, Niger Danube, Euphrates, Brahmaputra and Indus also belong to a bigger league. However, it is estimated by one source that the Ganga's full flood of 10,00,800 cubic feet a second is greater than the maximum discharge of the Misissipi.

Under the Ganga Action Plan, sewage has been singled out as the single biggest source of pollution in the river. But there are also 132 big industrial units and inumerable small ones. Very few of these have effluent-treatment facilities.

An abiding interest in the river has often resulted in strage, and sometimes unimaginable, parallels. For instance, victorian engineer, Sir Charles, Lyell, is reported as having estimated that 33,50,00,000 tons of silt gets discharged at Ghazipur and that this was nearly equivalent to the weight of 60 replicas of the pyramids.

Since the Ganga is revered, temples and scared ghats are intrinsic to its character. There are seven holy cities : Varanasi, Ujjain, Mathura, Ayodhya, Dwarka and Kanchipuram. At the Dasawamedh Ghat and in its immediate vicinity at Varanasi there are 692 temples.

It is ritual for bodies to be burnt at the ghats and for half-burnt corpses to find their way into the river. There is no one who has been able to hazard a guess at the number of such corpses, but one source has it that at least 200 bodies are burnt every day at the ghats of Varanasi.

The Ganga enters Bihar at Buxar and leaves at Rajmahal after covering a distance of 550 kms. The four class I cities Patna, Chhapra, Munger and Bhagalpur have been brought under the Ganga Action Plan, according to S.N P. Sinha, chairman-cum-managing director, Bihar State Water and Sewage Board.

In most parts of northern India, except for the brief cold season, the river water temperature is around 30° celsius. The study carried out by Bhagalpur University on the Bihar stretch of the river finds the same temperature range, most of the time, during the day hours. In the estuarine region of south Bengal, another phenomenon is observed. It is called 'vertical homothermy' which means that the temperature throughout the river is uniform; the temperature difference between the top surface and the river bed is 1.5°C (1967 71). The 200 km stretch of the river from Nabadwip to Kakdwip has maintained an unchanged temperature profile, ranging between 18°C and 33°C. In the Calcutta metropolitan area, the range is between 27°C and 33°C for most of the year. The river water temperature has a direct correlation with the sewage temperature of the region, because the same water after use takes the name of sewage.

When the temperature of water rises, the oxygendis soved in it difuses. One litre of water can hold at 0°C 14.63 mg of oxygen, compared to 7.5 mg at 30°C. The presence of oxygen water gives it power to combat the presence of oxygen consuming waste materials. The self-purification power of a river depends on the oxygen replenishment. Nature supplies according to her rules; artificial aeration can also induce oxygen to enter the water body. In fact, if the oxygen remains in surplus after oxidising all waste materials, only then is the self-purification process complete,

The biochemical oxygen demand (BOD) test, which originated in Britain, establishes the amount of oxygen on litre of water requires to oxidise degradable materials in it. The United Kingdom royal Commission on Sewage Disposal (1898-1915) found that the long term average summer temperature of British rivers was 18.3°C and their average flow time, from source to sea, was a little less than five days. It recommended that the BOD test in Britain would be carried out in laboratories at 20°C for five days condition after World War I; the UP adopted it in 1934.

British India, ignoring its location in the tropics, started measuring BOD according to British conditions too. The All India Institute of Hygiene and Public Health propagated it to the river boards. It was passed on to the Central Public Health Engineering Research Institute (now named National Environmental engineering Research Institute) and, finally, the Indian Standards Institution put its seal of approval to the unrealistic test conditions. Over 50 years, Indian scientists have collected BOD data which do not reflect the real condition. When pollution control boards were created, they insisted that the test conditions should be the same as in Britain.

Two foreign scholars who came to India to collect information water pollution control in hot climates were more than surprised, and politely disclosed that Nigeria, which was also under British rule, realistically measured BOD at 25°C. The World Health Organisation in tis report pointed out that BOD measured at 20°C for five days has the same value as at 30°C for the three days. Ten years ago, the Asian Institute of Technology at Bangkok concluded after a study that for southeast Asian countries, 30°C for three days conditions would be the most apropriate. But the Indian scientific community has remained non-responsive and dobious data is being collected, which cannot be the basis of any sewage treatment plant. Indian water is deficient in oxygen because of its temperature, which is higher than in cold countries, but this disadvantage is offset by the high temperature in another way despite the fact that the high temperature of water in India is a deterrent to the rentention of oxygen.

Ganga Action Plan : An Overview

The Ganga is India's pride. Since it flows through so many acient and important centres like Allahabad, Varanasi and Patna, the Ganga also symbolises our culture and civilisation. The river has tremendous potential but that is not being exploited properly. In European countries rivers are used so well for navigation, transporation and irrigation. We have not been able to do the same with the Ganga.

Besides, there is so much filth that flows with the river. Looking at the Ganga from the broader perspective of ecological balance and health, the 'Clean Ganga operation' is a very positive step. A filthy Ganga has a definite impact on our people because millions drink its water daily. Since 80% of all diseases are water borne, it is of utmost importance that we keep the Ganga clean. All states in the Ganga basin should endeavour to keep the Ganga Pollution free. In this context, the Ganga Action Plan undertaken by the centre is a very worthwhile project. However, the Ganga Action Plan will take some time to be implemented completely. No one has done anything to clean the Ganga until now. Factories have been built on both banks of the river since the days of the British. This continued even after independence and, unfortunately, though there were antipollution laws, nobody cared to enforce them. It is regrettable that no effective steps were taken earlier to clean the Ganga, which is of utmost importance in our country. By formulting 'Ganga Action Plan, the centre has for the first time taken an initiative to clean it. The people should also cooperate with the government to keep the Ganga clean. Unless a genuine awarness is created among the masses, mere laws cannot do all that we want to keep the Ganga clean.